# Life as Energy

Masterpiece of a brilliant young mind, *Life as Energy* exquisitely illustrates the elemental links between the hidden dynamics of energy physics, the connecting creativity of our planet's living web, and the unhappy yearnings of our now profoundly out-of-place human lives. Advancing the vision of such masterworks as James Lovelock's *Gaia* and D'arcy Thompson's *On Growth and Form*, the world needs more of Dr. Pietak's work soon.

— Dr. S. J. Goerner, Director of The Integral Science Institute, author of *After the Clockwork Universe.*

Life as Energy is a wide ranging, well written, and readable book drawing important parallels between physics concepts such as energy and quantum mechanics system states and the characteristics of complex living systems. The result is a novel framework that helps make sense of biological life cycles, ecology, and sustainable interactions with our environments. I am most impressed by the clarity and breadth of the many concepts Life as Energy explains at a level accessible to a general reader.

— Michael Sayer, Professor Emeritus of Physics at Queen's University, Canada

# Life as Energy

## Opening the Mind to a New Science of Life

Alexis Mari Pietak

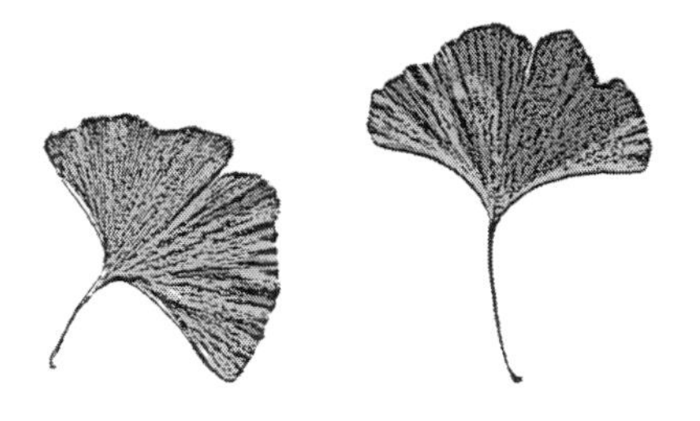



Floris Books

First published by Floris Books 2011



British Library CIP Data available
ISBN 978-086315-797-4
Printed in Poland

# Contents

To *La Que Sabe* in each and every one of us

## *Acknowledgments*

*I thrive best hermit style*
*With a beard, and a pipe,*
*And a parrot on each side*
*But now I find*
*I can't do this without you*
**~ Björk**

I'd like to thank the Universe for bringing this through me — it's been an extraordinary experience. Thank you to my mom, Dale Pietak, and my pa, Stanley Pietak, for a lot of love and support. Thanks to Liz Brehmin and Kate Jenkins for your supportive friendship. Thank you to Cory Dean, Andrew Shaw, and Michael Lewis for the deliciously rich and intense conversations that fed these ideas and made them strong and vital. Thank you to Andy Love for sharing the joys, trials and tribulations of a creative journey's beginning with me. Thanks to Andrew May for constructive feedback on the early stages of this book. Many thanks to Professor Michael Sayer, trusted Physicist and friend, for editing, input, and advice. Thank you to Christopher Moore for many helpful, insightful comments that shaped the final nature of this work. Thank you to all the staff at Floris Books for making this work possible. Thanks to Sess Curry, my beloved, loving companion — for your beautiful presence and support I am immensely grateful.

# Introduction: Singing the Green

*We believe our scientific view will never change because it is too well demonstrated. At the same time, we also realize at some level that this same science is killing us. It is powerful but empty and cold.*

**~ Sally J. Goerner (*After the Clockwork Universe*)**

*To sing means to use one's soul voice. It means to say on the breath the truth of one's power and one's need, to breathe soul over the thing that is of one's power and one's need, to breathe soul over the thing that is ailing or in need of restoration.*

**~ Clarissa Pinkola Estes (*Women Who Run With the Wolves*)**

There is an ineffable, yet significant and intuitively recognizable difference between those things that are alive, and those that are not. Humanity's older beliefs, and the colloquial language still used today, refer to life as a *force* or *energy*, no less real than the grounding force of gravity, the warming energy of the sun, or the powerful flow of a rushing river. Life energy is the *qi* of Chinese medicine and the *prana* of East Indian Ayurvedic systems. It's seen as an inexhaustible ordering influence. We know that a living thing holds together a pattern, grows, develops, heals, reaches out to, recoils from, and may feel, think, choose, and be aware. We speak of a living thing as something in possession of this property called *life*. In turn, the loss of life is consistent with the loss of a property, as the pattern once maintained falls apart, once moving becomes motionless, once warm cools, and once responsive, is no longer so. On an intuitive level, the differences between the living and the non-living are profound and rarely mistakable; however, the idea of life-as-energy remains excluded from scientific systems of thought. In the eyes of science, there is no significant difference between the living and the non-living.

Energy is a property used extensively in the sciences. Yet, like life, it's not possible to break the concept down into more basic, explanatory ideas. Quite simply, energy is a property that makes something happen or stops something from happening. In other words, energy is the ultimate agent of causation in the universe. We understand that an object is warm because the molecules that make up the object are moving, and that the molecules are moving because they possess energy. We understand the river's flow to be impeded by a dam, as the dam is composed of bound molecules that will not break until the energy of the river's water exceeds the energy of those bonds. Intuitively, it is not difficult to extend these concepts of energy to account for the unique properties associated with living things. Energy is widely accepted by science as a property of the universe; however, the idea of life as a property, of life as an energy, is ignored and rejected.

Although many of us may be excluded and/or indifferent to science and its role in society, it is immensely important as a deciding force for what is real, sensible, and worthwhile. Look around at the many gadgets, medicines, and strategies that presently shape the character of our lives and even the very structure and function of the planet. Of course, westernized cultures largely base their activities and design their tools on knowledge that is derived and endorsed by science. Therefore, science shapes the tangible structures of our human lives and worlds. Science also helps cultivate the relationship between us and the myriad of living things, materials, and forces surrounding us. In short, science decides between what is real and what is fanciful. It is therefore very important what science thinks about life.

Our sciences, in particular the so-called life sciences (medicine, botany, cellular biology, and so on), insist that living things can only be understood in terms of their small material parts (cells, genes, molecules) and that there is nothing particularly special needed to differentiate a living from a non-living thing. In this modern view, you and I are reduced to chemical machines functioning on the software programs of our inherited genes. Our physical diseases, mental illnesses, and even emotions like love, are reduced to the presence or absence of molecules in the chemical bags of our bodies. A doctor ascribing to the modern view of life will only want to give you the newest pill, side-effects and all, insisting it is the only viable treatment. The depressed, anxious, and otherwise mentally-ill are rarely seen within the greater context of their personal lives, or the context of their society, but rather, as those suffering from a chemical imbalance requiring the administration of another chemical treatment.

Although this modern view of life has achieved many remarkable successes, it's also associated with significant drawbacks and limitations. Notably, since science tells us there's nothing particularly special about life, it's easy to adopt the notion that the beings surrounding us are essentially objects, and as objects, can be justifiably treated as such. For instance, much of 'life science' research takes living, healthy animals to be not much different from usable, dispensable objects. In universities where I have worked, researchers have required defects in bone that were too big to heal on their own. For the living creatures they tested on (rats, rabbits, pigs, sheep) this meant removing the greater portion of their skulls to leave their brains exposed, or a large portion of their jaw or arm bone, leaving the living animals in this condition for up to twelve weeks afterwards. In another research group, rats are choked — but not to death, just to dysfunction — in order to create brain damage which they then attempt to remedy.

In addition, the modern view of life teaches us to think that a complex phenomenon arises because of a single factor existing in the tiniest parts making up that phenomenon. Also, something is typically studied by extracting it from the complex context of its natural environment to the simplified, controlled settings of a scientific laboratory. Both one-factor/one-phenomenon thinking and the simplification of a phenomenon's context are bad thinking habits that can be seen to contribute to the numerous, serious, often globally-scaled, long-term side-effects which occur when our technologies (the application of scientific knowledge) are re-introduced to the complex, interconnected environments of whole organisms, whole ecosystems, and whole planets.

However, this modern view of life was not always the prevailing scientific perspective. For instance, at the turn of the twentieth century (between about 1910–1940) scientific ideas of life that allowed for its wholeness, and were capable of entertaining its exceptional qualities, were budding into existence. Early scientists who studied the development of a life form from a seed or egg saw the intricate order arising from a clump of selfsame starting material to be the result of an energy field (like a magnetic field), which was self-created by the biological system and served to inform and coordinate cells over a wide expanse of space and time. Some of these early life scientists even equated the fundamental character of this biological field to the energy of life itself. However, by the 1950s, genes, DNA, and the molecular mechanisms involved in many conditions were discovered. For a variety of circumstantial and political reasons, ideas such as the biological field were abandoned in their developing stages in favour of the modern view of living things. Perhaps it's assumed that there are sensible scientific reasons for the abandonment of these early ideas, however, to this day they remain neither refuted nor proven.

In contrast, at about the same time period (1920s onwards), the realm of the physical sciences came alive with many seemingly outrageous ideas — ideas that do not require a justification in terms of their itsy bitsy parts in order to be appreciated and used. In the physical sciences one finds well accepted concepts stated in terms of wholes and non-material based entities such as energy fields and quantum wavestates. Albert Einstein, perhaps the world's most famous scientific genius, showed us that matter was, in its essence, just bound-up energy, and that traveling very fast in a spaceship would cause time itself to slow down. In quantum physics, the behaviour of very tiny particles was found to depend on whether or not the scientist was 'watching' them or not! In addition, the entities of the quantum realm need to be interchangeably thought of as particles, waves, or energy fields, and this causes no issues for the physicist who can change their conceptual view of a phenomenon as we would change a pair of glasses.

It now seems as though the trajectories of the life and physical sciences are unique and irreversible, having occurred for very good reasons, and leaving each scientific field permanently set in its particular course and thinking style. However, in questioning how the life and physical sciences could take such divergent paths, we see that a particular style of thinking has been lost from the life sciences, but not the physical sciences. As we'll explore in this book, a less recognized thinking style blending imagination and rationality (*imaginative rationality*) can be extracted from the physical sciences and extended to the life sciences. This is a thinking style that demonstrates the concept of life-as-energy to be no less scientific than other well-accepted concepts in the physical sciences, and worthy of serious reconsideration. Imaginative rationality, which is used all the time in physics, is a viable alternative to the thinking habits that have taken over the life sciences.

In this book we'll begin by revisiting some older concepts of the early life sciences, namely the biological fields once thought to be involved in the development of a life form. From the perspective afforded by imaginative rationality, we'll see that biological fields are really no less 'scientific' or unfounded than other energy fields of physics. Yet, this is just the beginning, for we'll use imaginative rationality to develop and expand upon biological fields to arrive at a system of ideas that indicate what life-as-energy might mean, and what it might stand for in an individual life form. Finally, by merely allowing life energy to exist, we'll find that very complex processes such as the spontaneous *natural succession* of a plot of bare land through characteristic stages of vegetation (from grassland, to bush, to coniferous forest, to deciduous forest), as well as the life cycle of most organisms (development from a simple state like a cell, through to a mature organism, through to inevitable aging and death) can be easily accounted for. We see that by merely allowing life-as-energy to exist, we arrive at ideas with applications in ecology, agriculture, and care of the environment.

In the myths of many diverse groups of first peoples, the act of singing has been synonymous with the act of creation. The Aborigines of Australia, who perceive land itself to be alive, speak of ancestral beings who wandered about the continent singing birds, animals, plants, rocks, and waterholes into existence. Similarly, the Navajos of what is now the American Southwest tell a creation myth where the entire Earth arises from sacred chants sung by the first 'Holy People'. In Mexican mythology, we find the Wolf Woman, *La Loba*, an old crone living in the desert collecting the bones of all sorts of deceased creatures. La Loba is said to carry these bones back to her cave where they are carefully assembled. In singing over the bones, La Loba brings these creatures back to life.

Of course there are many different ways to create and to sing. Our 'modern', 'Western' civilization has been singing too. Yet the way we have been singing has sung up forms in asphalt-black, concrete-grey, and liver-brick-brown. We've been singing in all sorts of deadening tones. This singing we're referring to represents the underlying thinking styles that shape our very perception of the world around us, and thereby influence how we will act and create in that perceived world. *How* we think is a direct mirror of the relationships we'll forge, and the opportunities, problems and solutions we'll perceive. Therefore, to change our actions and creations, we must learn to change this very basis of how we are thinking; of how we are singing. It is now time to sing in a different way. 'Singing the green' is to sing humanity as one type of cell in the interconnected living body of the Earth; is to sing back her florid forests and her clean rivers; her estuaries and elephants; her birdsong and bumblebees; her hummingbirds and humus; her oceans full of life. It is to sing her our mother, our lover Earth. It is to sing another world altogether, by changing the very way that we think about it. Since science defines and underlies our culture's rationality, it is in part through science that these changes must come. It is time to breathe soul over the life sciences to open and restore them for the sake of all of life and things living.

## *Summary*

I like to think of this book as a journey — a thought journey — that you, the reader, and I, the writer, are going on together. The major aim of this journey is to demonstrate that merely the recognition of life-as-energy supplies a practical and comprehensive context through which some outstanding properties and processes of living systems can be accounted for. However, this book is also intended to raise your personal awareness of what makes something scientific, how science is conducted, the mindsets that modern science is struggling to work beyond, and most importantly, the way to authentically bring more imagination and open-mindedness into scientific thinking styles to move beyond the cynical, clinical view without compromising on the fundamental 'rational' basis of science. Thus, the first chapter of this book is a look at the importance of science in deciding what is real, effectual, worthwhile and rational on our behalf. In Chapter 1 we will also consider the fundamental mindsets of the past few hundred years of 'modern', 'Western' science and consider how these have cultivated a particular type of relationship between scientists and their objects/subjects/beings of study. We'll consider specifically how these particular styles of thinking may be limited, especially in their dealings with living things. We will also consider alternative scientific perspectives of living things that have been cast-aside in the past few decades.

In Chapter 2 we'll look in more detail at the many unique and perplexing features of living systems so we can re-appreciate their awe inspiring features, and also be able to understand the aspects of life and things living that confound the mainstream mindset of the modern life sciences.

In Chapter 3 we will consider one alternative scientific thinking style based on wholes and imagination, which has been used to develop various fundamental and well accepted concepts in the physical sciences. We'll consider how this 'imaginative rationality' may be used in scientific comprehension of life and things living. We'll re-examine the more holistic concepts such as life-as-energy and biological fields (morphogenetic fields) as new and useful contexts in which we can work with living things.

In Chapter 4, portions of science that are drawn upon in future discussions of life energy using imaginative rationality are described in a visually based, maths-free manner that is intended to be accessible to those readers who may not have much background knowledge of the sciences.

In Chapters 5 and 6, the qualities, processes, implications and applications of life energy are developed using imaginative rationality. Having come to understand underlying processes by which scientific concepts are generated, and having explored some basic concepts of science from this perspective, these well founded scientific systems of thought are used to engender new conceptualizations and understandings of things of a living nature.

Chapter 7 is an exploration of some mainstream human activities from the new perspectives developed in the preceding chapters. Alternative perspectives and solutions in the realms of ecology, agriculture and healthcare are discussed.

# 1. Ghosts in Our Oceans of Thought

*The image of the world changed to that of a machine and the ambition of science was to dominate and conquer Nature. Such an entirely material world could be treated as if it was dead, letting man be the possessor and master of his environment, including all plants and animals ... all the mysteries of Nature could now ultimately be explained in mechanistic terms.*
**~ Lars Skyttner (from *General Systems Theory*)**

## *Science matters*

*Science and engineering have been unable to keep pace with the second order effects produced by their first order victories.*
**~ Gerald Weinberg**

A lot of trust is put into scientific knowledge and its technological fruits. Concepts that are scientifically proven are made credible, and this can make them quite valuable in their application as technology. Suppose for instance that someone were to claim that an extract of the fungus growing on an orange peel could cure life-threatening infections like pneumonia. If scientific studies support the validity of this claim, then extracts of this fungus will likely become a new medicine with widespread use and acceptance, as has been the case for the antibiotic penicillin, derived from the fungus penicillium. However, if scientific investigations uncover no beneficial effects, then this claim isn't likely to be given any credibility, and a new medicine is very unlikely to be developed. From science the human race has inherited numerous lifesaving medicines and medical procedures; gadgets and tools useful for enhanced food production; convenient transportation and engaging entertainment; various eating and lifestyle strategies for optimum health and safety; implements like nuclear warheads for unthinkable mass destruction; and a rather astounding understanding of impossibly far away wonders such as galaxies, suns, black-holes, and the tiniest bits of quantum matter leaping in and out of existence. As debates on climate change will attest to, we also count on scientists to determine the reality of, and the solutions to, various environmental issues. Ultimately, science matters so much because in modern cultures it defines what is rational. Rationality implies what is sane, reasonable, worthwhile and safe, and distinguishes this from what is crazy, nonsense, worthless, and even dangerous. Many modern cultures have decided it best to act on this 'rational' basis, and thus, scientific knowledge has become the platform and directing influence behind many cultural activities.

Science is important as a primary deciding factor of what is rational; however, to a large extent, the people, knowledge and methodologies of science remain fairly inaccessible to the average person. For the most part, scientific investigations are conducted in institutionally-based laboratories locked away from the mainstream world. Scientific communications are so inextricably laced with technical terminology and densely imbedded concepts that formal academic training is usually required to decipher its reports. Moreover, much of scientific knowledge is published in closed-access academic journals that require an affiliation with a university or other subscribing institution in order to merely make the scientific knowledge-base and its recent developments available. Science has such an integral role in directing cultural activities and developments, and yet, due to its inaccessibility, an implicit trust or a blind rejection of science are the general options available to the public at large. Perhaps as a result, many people don't seem to care much about science, content to let it spin away in its isolated laboratories to occasionally spit out a new gadget or notion of interest. However, it seems to be taken for granted how insidiously science has infiltrated into secular thinking styles, perceptions, choices and activities. To be largely unaware of science and its processes is to let it have the best opportunity to control and decide what is 'rational' on your behalf.

The past half century has been marked by various unexpected events and ecological catastrophes that have often occurred in the wake of various technologies (the direct applications of scientific knowledge). These assorted crises imply that there are significant limitations, devastating shortcomings, and large blind spots in our trusted sciences. It therefore seems ever more important that we take an interest in the fundamental nature of science and what it's cooking up at any point in time. A dissemination of the foundation from which science tends to operate may allow those without academic training to make informed choices about what ideas and activities they discern to be rational, and those they feel are misplaced or even insane, independent of their endorsement by the modern scientific community. This is especially important in this day and age as science encroaches further and further into a disenchanting parameterization and manipulation of living things. Meanwhile, various New Age spiritual doctrines offer up alternative viewpoints and strategies to those posited by science; viewpoints which may be economically lucrative for gurus and practitioners, but may or may not be viable interpretations of reality. Therefore, in this chapter we begin our journey by asking and exploring what it is that science intends to do, and how a scientific investigation is conducted. We'll pay special attention to the assumptions of science, and to the perspective of living things that has evolved from the world-view created by these assumptions. We will see how the cynical, clinical perspective of life and things living has evolved in Western sciences to become the dominant perspective of this modern age.

## *The buried beliefs of science*

*Reality is that which, when you stop believing in it, does not go away.*
~ **Philip K. Dick**

Scientific knowledge is derived from a community of human beings who have agreed upon a process for investigating reality, and who have clear and specific aims for deciding what constitutes an acceptable answer to these investigations. Scientists tend to make a number of assumptions about the fundamental nature of the universe as perceived by us human beings. These assumptions arise from underlying beliefs that influence the questions scientists will ask, and the specific methods that science will use to answer these questions. These all cultivate a certain kind of relationship between scientists and their subjects of study. At the root of all of this is a basic model for human experience that *appears* to set the stage for how science must ultimately function.

Let's begin at these roots by considering this model for human experience that divides the world into two vantage points formally called the *objective* (the world out there) and the *subjective* (the 'place' we'd call your mind, self or imagination). In this familiar model we sit within our bodies looking out to the objective and looking into our subjective realities. We are simultaneously ghosts in the formless internal ocean of our subjective thoughts and feelings, as well as corporeal participants in the more solid and more real objective place of material forms and events. Some, but it seems not all, of our formless internal processes correctly interpret the fundamental nature of the objective world out there. We call this ability to access the 'true' nature of the objective realm our sense of *reason* or *rationality*, and the portion of our subjective experience that accurately corresponds to the objective world as that which is *reasonable* or *rational*.

Science can be defined as a process for making the most of our sense of reason in order to determine the true nature of the one objective reality, free from any contaminating aspects of our subjective inner worlds. These subjective contaminants are generally our feelings, and beliefs based on our feelings. The notion of subjective feelings and beliefs as contaminants arises from the sense that I can wish for or want something very strongly, but am not able to get something to happen by wishing and wanting on its own. I may really want a rock to lift off the ground, but cannot move the rock from its particular place by imagining alone. Science is diametrically opposed to wishful thinking. Science sees it absolutely necessary to conduct a controlled experiment to ensure that what the scientist feels or believes that the outcome should be does not influence the results. Thus, 'scientific' is often synonymous with emotional detachment. Unfortunately, as relationships most often involve some emotional attachment or at least some emotional aspects, the side-effect of emotional detachment tends to be a sterilization of the relationship between a scientist and their object/ subject/beings of study.

An objective fact is typically a perception of the world-out-there that remains fairly constant and matches-up between people. Objective facts most often involve matter, the stuff of the universe. If for instance, a dead bird lies on the ground, many different people may observe and verify the fact of the dead bird on the ground. Observations of the dead bird are therefore easily ascribed to this so-called objective reality. Yet, how the different people *feel* about the dead bird, or what the different people *believe* to be the reason that the bird is dead, may vary significantly. These beliefs and feelings about the situation are ascribed to the subjective reality of these individuals. A scientific investigation of this situation would aim to determine what has happened in objective reality to cause the bird's death: the objective fact of the situation.

More explicitly, the *scientific method* is (usually or supposed to be) based a cyclic set of steps that assist in uncovering the uncontaminated objective truth of a situation or phenomenon. The scientific method begins with some initial observation of a phenomenon (for instance, the dead bird on the ground), which engages some reason-endowed person to formulate a question about the nature of that phenomenon's occurrence (why is the bird dead?). This scientific observer, after giving the situation some thought, then proposes a first possible, hypothetical answer to the question, which is aptly called the *hypothesis* (the bird ate poison). Then, in order to test the hypothesis, an experiment is conducted to investigate the phenomenon under controlled conditions while eliminating any personal bias of the observer's beliefs and feelings which might unduly influence the outcome of the investigation. This requires, among other things, that no observations of the experiment be thrown away or manipulated in any way by the scientist. Thus, the hypothesis is tested by acquiring further neutral observations of the situation, or by recreating and observing the situation under controlled conditions to generate data which are referred to as the *results* (in this example the bird might be dissected finding not poison but a crushed skull). These results are interpreted using the scientist's inherent faculty of reason (the bird died not by poison but by flying into a nearby window). This interpretation in turn becomes a new hypothesis which ideally leads to a further experiment, results and interpretations. This iterative process is intended to map out some truth or regularity of the universe, and is the formal process of science. The outcomes of this process (which generally continue to evolve) are what we call *scientific knowledge.*

This model for human experience as a ghost imbedded in both a subjective ocean and as participant in a singular objective reality is a very nice, intuitive picture, which forming the basis of scientific perspective has assisted with determining a number of consistent things about the universe. However, working from this model, someone may become accustomed to perceiving reality in this certain way, become forgetful that this is but one perception of reality, and come to be limited by the base model. For instance, working from this model a scientist may insist on the ability to obtain some absolute truths about objective reality, and may consider those truths to be transcendental of human experience. With this notion in mind, they may write their embodied human selves out of the equation, and in doing so, become unconscious of the fact that there even are underlying beliefs held by the scientific regime itself. Yet, as we'll get to in a moment, these underlying beliefs do exist and, especially as unconscious assumptions, are very important factors shaping the very nature of science itself.

The underlying beliefs of science have a great influence on how science will operate, as to a great extent they influence the perception of reality, the questions that tend to be asked within the context established by the perception of reality, the methods used to answer these questions, and even what possible answers are forbidden. These underlying beliefs determine the kind of relationship the human scientist will have with their subject of study. For instance, in the seventeenth century, René Descartes, an influential philosopher and early scientist who became a prototypal figure for much of modern Western science, viewed the entire universe as a large machine composed of parts functioning together like the parts of a clock. Descartes' machine view extended to include non-human living things such as dogs, cats, and birds.[1] Seen as directly analogous to a machine, these creatures became objects believed incapable of feeling pain, emotion, or other sentient qualities. Descartes' perspective of the world was widely accepted and grew to be part of an influential scientific movement in Europe. Yet, this fundamental belief of animals as machines promoted the practice of *vivisection* — the dissection and subsequent experimentation upon animals while alive and fully conscious. In the name of scientific knowledge, the legs of healthy, conscious dogs were nailed to floorboards, and with a scalpel the dog's abdomen was sliced fully open. Great care was taken to not mortally wound the creatures in the process, allowing the scientists to see physiological processes such as the beating heart, and to understand its role in blood circulation. The tremendous cries and yelps of the living dogs were perceived to be no different than the clicking and rattling of the parts of a machine. In the modern age, in most developed countries, animals such as cats, dogs and primates are no longer believed to be entirely non-sentient machines, and the relationship between the scientist and their beings of study has changed, at least to a matter of degree, as anaesthetics and analgesics are now mandatory for (most) animal experimentation that causes more than momentary pain.[2] A degree of change in underlying belief systems lead to a degree of change in the relationship between human scientists and their subjects of study.

Recently, many of us may have noticed that our human technologies, in particular the relationship between our technologies and living things, have been associated with unexpected and often catastrophic outcomes.[3] These unexpected outcomes indicate that the scientific reasoning from which these technologies arose, and that our culture has come to trust so implicitly, is perhaps more irrational and dysfunctional than initially assumed. Chemicals such as DDT (*dichloro-diphenyl-trichloroethane*), designed to kill insects, have concentrated in food chains, significantly killing off birds and mammals at the top of the food chain; commonly used chemicals now present in drinking water and even found as contaminants in human breast milk have been found to cause cancer and birth defects; carbon dioxide from machine and factory exhaust has built up in the atmosphere, contributing to climate change; stem cell therapies, promising healing in many degenerative conditions, have recently been implicated in causing cancer. While there are a number of socio-economic and other factors at play in these scenarios, at the least they reflect a lack of awareness of the complex, interconnected nature of environments in which technologies ultimately participate. Taking incidents like these into account, it is clear that our present, mainstream scientific regimes make mistakes, and that these mistakes may very well be related to their focus on isolated processes occurring in simplified environments. The dysfunction of their approach shows up when the technology is introduced to the natural system (body or environment) from which the phenomenon was originally isolated. This focus on isolated material parts functioning in simplified environments is the legacy of particular styles of thinking that pervade the sciences even to this day; tightly interwoven thinking styles which go by the names of *reductionism*, *mechanism*, and *materialism*.[4]

## *The rise of scientific mindset: a tale of three -isms*

*It is the absence of any theory of organisms as distinctive entities in their own right, with a characteristic type of dynamic order and organization that has resulted in their disappearance from the basic conceptual structure of modern biology. They have succumbed to the onslaught of an overwhelming molecular reductionism.*
**~ Brian Goodwin (from *How the Leopard Changed its Spots*)**

When we find a way of thinking that supports a way of acting that helps us to function more effectively in the world, we'll probably try it again as it has worked before. Perhaps the outcome will again be favourable and this will strengthen confidence in the particular way of thinking. Then, when a similar situation comes along, we may think that way about it again. The more we think a certain way, the more solid and set the thinking becomes. Eventually, it becomes a *mindset* — a way of thinking that is seen as the only way to approach the problem. Eventually it *is* the way things are. A mindset is a problem because it is so sure of its rightness that it rarely questions itself, and when mistakenly used in situations perceived to be similar to the one where it worked well in, but which are woefully dissimilar, the results can be devastating. Conditioned to believe this is the way that things are, the mindset persists, and may attempt to carry out its objectives with more and more vigour, determined to make things work. This seems to have happened in the sciences, in particular the life sciences, where a particular style of thinking continues to be widespread in spite of mounting evidence of its limitations and inadequacies.[5]

An experience of the world can be whole — a full, unbroken continuum of pressure, light and sound. Yet when mind emerges with the process of thinking, the substances of experience are separated, named, and defined. With thinking, one thing is differentiated from another type of thing: this is form and this is formless, this is colour and this is sound, this is me and this is not me. Human thinking brings with it language, naming, and the use of categories as mental containers to hold concepts of related qualities. There is colour, and then there are different kinds of colours. The plant has leaves and flowers, and the leaves have veins and the flowers have petals and stamens and powdery pollen, and yet all belong to the form of the plant. One predominant, inherent quality of thinking is that it resolves, separates, reduces, focusses and recollects various aspects of the world. Perhaps not surprisingly then, the platform of science tends to rest on habits of thinking that are rooted in dissecting material reality into smaller and smaller elements on to which focus is maintained before they are re-integrated and related back to the whole from which they came.

Thinking styles that have significantly influenced today's 'Western' sciences can be traced to ancient Greek philosophers who lived and thought approximately 2400 years ago. Among other things, these early thinkers (such as Democritus, Aristotle) developed the notion that a physical thing (such as a crystal), phenomenon (water turning into ice crystals), or even an idea could be *reduced* into a set of key elements (parts, steps, or axioms). It is assumed that by understanding the nature of these (presumably simpler) elements, the properties and behaviour of the complex whole could be deduced. This approach has been aptly called *reductionism*. In a reductionist perspective, the properties, processes, interactions and relationships between the parts are what generate the properties and phenomena witnessed in the whole — not the other way around. This places a special import on the parts as, if they are the agents that cause things to happen, then knowing about them can be used to gain predictive or manipulative control over the phenomena presented by the whole. This way of thinking of the parts as causing the properties of the whole is formally called *upwards causation* and is quite implicitly associated with a reductionist thinking style.

To intuit how the reductionist style works, let's consider how the basic parts of matter (the 'stuff' of the universe) can be seen to account for the properties of matter as we observe it in our everyday human worlds. Matter is commonly observed in distinguishable forms that we call solids, liquids and gases. Remarkably, as far back as 450 BC, the Greek thinker Democritus rather accurately postulated matter to be composed of fundamental units which he called *atoms*. Today we have found ample evidence that atoms actually do exist, and while they contain further structural details, atoms (or molecules, which are clusters of bound atoms) can be seen to account for some properties of matter. For example, different types of relationships between the same kind of atoms/molecules translate into the distinguishable properties of the various phases of matter (as implied by Figure 1.1). Firm solids such as ice represent tight, highly organized arrangements of molecules; flowing liquids like liquid water represent more loosely interacting molecules with little structural organization; and ephemeral gases like water vapour represent a case of almost entirely non-interacting molecules. In all three cases, the parts are the same water molecules, but different organizations and relationships between these same water molecules account for the different properties of the whole. Thus, the explanatory power of the parts (atoms or molecules) in determining the behaviour and properties of a whole (phases of matter) is rather clear.

A reductionist thinking style is implicit in the design of human machines such as clocks, automobiles, and spinning-wheels. A machine is made of clearly identifiable parts, each performing specific functions that result in a functional whole. While reductionism brings us the notion that a whole can be understood in terms of its resolvable parts, *mechanism* adds that the behaviours of these parts are governed by

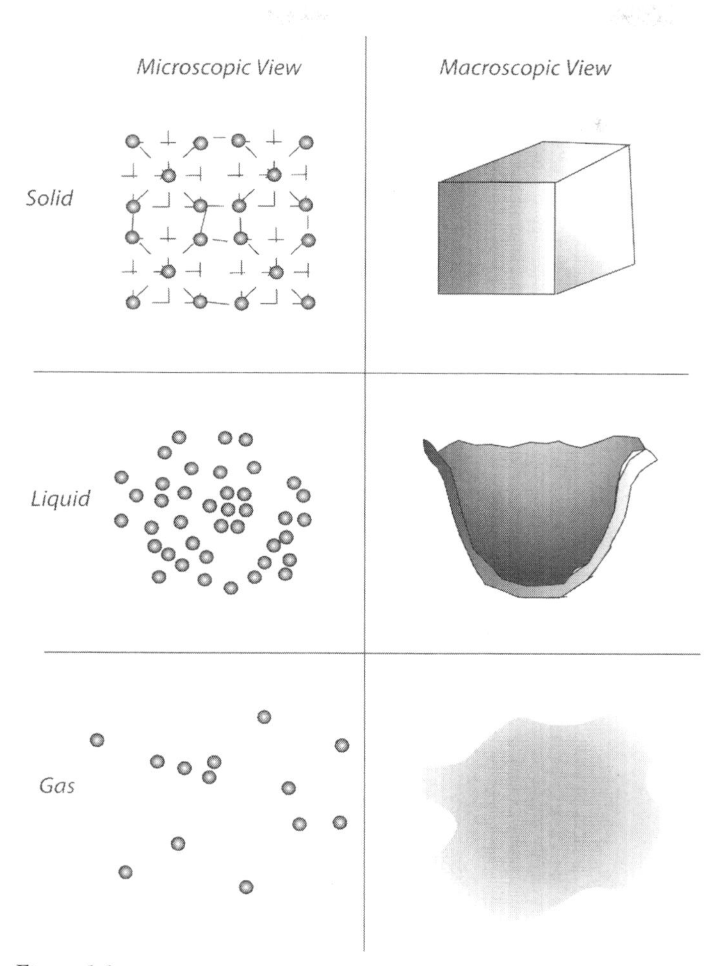


*Figure 1.1*

consistent universal laws which apply to all material forms in the universe. Therefore, the behaviour of the whole, in terms of its parts, can be determined by knowing the parts and the laws pertaining to the parts. These governing mechanisms typically originate in the physical sciences (physics) or the chemical sciences (chemistry) which are often seen as the foundation of science. Thus, a mechanist would ultimately affirm that living things would ultimately be described in terms of known physical and chemical laws.

Closely associated with reductionism and mechanism is *materialism*, which is to focus on the material based stuff at hand as it is perceived to be the only thing to truly exist. To someone adhering strictly to materialistic viewpoint, something like love or an idea does not actually exist as it's not made up of atoms and molecules. To a materialist, love and thoughts 'exist' only in our heads; they are part of subjective reality and take secondary importance to the material stuff of objective reality.

These three -isms of reductionism, mechanism and materialism are interwoven thinking styles that form the spectacles through which much of conventional science views reality. Seen through these spectacles, the existing things of the universe are made of matter, and are also inherently machine-like, consisting of clearly identifiable material parts that act under the governance of universal law to form a functional whole exhibiting various determinable properties. Therefore, to understand these properties, the material of the whole must be broken down into its basic building blocks. These are the prime agents that cause things to happen in the whole, and are typically atoms or molecules. Once the parts are resolved and the laws governing their behaviour are determined, they can be related to the properties of the whole.

The reductionist, mechanistic and materialist thinking habits were, and continue to be, applied to living things. From these endeavours it is expected to obtain a functional understanding allowing for prediction, manipulation, and control of the properties of the whole by acting on its parts. Admittedly, it is actually not hard to see a living thing in terms of the three -isms perspective. Upon simple examination, a living thing is indeed found to be composed of various clearly identifiable material parts. At first glance these are the organs (heart, liver, kidney, and so on), which can be seen to function together like the parts of a machine. In turn, these organs can be each seen as individual machines composed of clearly identifiable cells that perform various functions for the organ. Different cells in turn appear to be like little beings unto themselves, with a resolvable living-dead state somewhat independent of the life of the whole organism, and clearly resolvable *organelles* akin to miniature organs. Inside the cell there are the clearly identifiable *mitochondria*, which are like batteries preparing energy for the cell from sugar and other foods; and the *golgi apparatus*, which are like miniature factories preparing and processing various cell products. Under the three -isms view we keep on going like this until we see the individual molecules participating in the individual processes of the cell organelles, and trust that by understanding things at this level we can work our way back up again to account for the properties of the whole organism.

When applied to living things, the conventional protocol indicated by the three -isms appears to work quite well. If for instance the whole organism has a problem assimilating sugar for energy (a condition called type I diabetes) by looking to the material parts of the living 'machine', a single isolatable molecule called *insulin* will be found deficient in this condition. By supplying the living thing with more insulin, the condition is treated. In this case the three -isms are supported as a view for living things as it is possible to manipulate the phenomenon exhibited by the whole (diabetic condition) by acting on just one of its material parts that is the prime causative factor (insulin molecules). As another example, thoughts and emotional states can be seen to arise from the activity of the brain. The brain is composed of neurons interacting with various chemicals. Therefore, the activity of neurons and relevant chemicals has been associated with thinking and states of mind and emotion. From the lens of the three -isms, depression is seen to be caused by low brain levels of a molecule called serotonin, and can therefore be corrected by partaking of a molecular medicine (Prozac or other alternatives) which increases serotonin or its activity in the brain. As this strategy has been scientifically proven to be effective in treating depression, the thinking style of the three -isms is again vindicated. However, the side-effects of these treatments and their success in the long-term tend to be ignored and understated. As we'll see in the next section, the three -isms perspective ultimately becomes problematic and limited as it loses sight of the larger context and interconnected nature with which the phenomenon is truly associated.

Like any mindset, as the lens of the three -isms appears to work when applied as a thinking style to interpret living things, it has continued to be applied in many diverse situations involving many phenomenon exhibited by living things. However, like many thinking styles that have met with some success, the three -isms have become a predominant scientific mindset, rarely questioning its base assumptions, ignoring outlying factors which don't fit into the scope of the thinking strategy, and often rejecting alternative perspectives that are not based on its tenants. In personal communication, a renowned, predominant biologist maintained '*in my view, any model that describes biological phenomena has to be in principle interpretable in molecular terms*', thereby automatically dismissing any theories of life and things living that were not explicitly stated in terms of, or in reference to, molecules. This perspective reinforces the three -isms thinking style within the academic community, pushing down new sprouts of alternative, holistic views that may be trying to emerge.

## *How living things confound the -isms*

Reductionism comes with the implicit notion of upwards causation, which places a special emphasis on the parts of a system. A primary difficulty in working with living things in terms of their parts is their sheer complexity. For instance, the average human body is estimated to have on the order of 100 trillion (100,000,000,000,000!) cells.[6] If you were to try and count these cells, and you counted at the reasonable rate of about 1 cell each second, it would take you just over three million years to count them all! Each of those cells engages in hundreds of thousands of potentially meaningful interactions as it participates in various biochemical, macromolecular, and physical processes to establish the properties of the whole. In addition to this sheer complexity, the details of what are happening in each of these interactions must be elucidated. Due to the immensely large number of parts and processes that must be considered and managed, even with today's incredible computers it remains extremely computationally difficult, if not impossible, to account for the global properties of the average living thing in terms of its parts. This task has, however, not swayed scientists, although in attempting to sort out these details, they end up drowning in oceans of data.

Though the organized complexity of a life form may be daunting, and make an approach via the three -isms difficult, this doesn't actually indicate a flaw in the underlying reasoning. Taking a wider view of living things, it does however seem that restricting the perspective to one of upwards causation is an overly simplistic and not entirely realistic interpretation of many aspects of living things. This is because in addition to their sheer complexity, living systems exhibit *emergent properties.* Emergent properties are confounding phenomena as their existence depends on an underlying substratum, such as particular organization of cells, but they are not to be found in any isolated part of that same substratum. Rather, these properties *emerge* from the dynamic interactions and relationships between the many individual parts of the complex whole to become something that is greater than the collection of parts. Therefore, an emergent property is irreducible. Thought is an emergent property. A thought cannot be isolated to a single neuron, but is the result of the collective, interactive activities of a great number of neurons generating long-range patterns of interconnectivity. Yet thought isn't merely the firing pattern of neurons either — it's an entirely new property with aspects unique to itself.

In addition to their irreducibility to the parts of the system, emergent phenomena can make things happen. This is indeed a confounding notion, as it suggests that emergent properties simultaneously arise from the actions of a collection of individual elements, and paradoxically, may also influence or even dominate the actions of the same collection of elements! As a simple example, my thinking arises from the activity of neurons, which are in turn supported by the other tissue and cell systems of my body. However, if I'm hungry and think of eating an apple, my body begins to salivate; if I think of something particularly stressful, my heartbeat increases and my hands turn white. My thoughts have a distinct causal effect on the material of my body, which in turn supports my thoughts. This ability of the emergent property to influence the lower-order parts from which it itself arises was described as *cyclic causality* by once of this century's great emergence scientists, Herman Haken.[7] This cyclic causality occurs in distinct contrast to the heavily assumed upwards causation of reductionist perspectives.

Emergent properties and cyclic causality highlight the immense importance of *context.* Effective work with emergent properties ultimately requires the definition of a new context. As an example, imagine attempting to describe a musical performance in terms of the air molecules that support the manifestation of sound. Ultimately, sound does arise from dynamic pressure disturbances propagating through a very large collection of air molecules. Yet music is much more effectively and practically handled by considering it an emergent phenomenon with properties and processes of its own accord belonging to an entirely new context. It is much more relevant and practical to speak of the tempo, rhythm, pitch and mood of music, rather than the kinetic energy, momentum and position of an immense number of air molecules as a function of space and time. The key point is that both the reductionist and holistic perspectives are viable and necessary, depending on the situation they're required to assist with. A reductionist view of music in terms of air molecules and sound waves is necessary for designing a theatre or an instrument. A holistic view of music in terms of tones and tempo is indispensable for performing songs. It's a similar case for the science of living things; however, at present holistic views are being dismissed and remain undeveloped in favour of three -isms perspectives.

Thus, we can take note of some technical difficulties and confounding properties of living things which shake the basis of the three -isms: their complexity resulting from a large number of intricately organized parts and processes, their emergent properties and cyclic causality, and the need for new context in which to work with emergent properties.

## *Complex systems: evolving the three -isms perspective*

In recent years, the three -isms perspective has evolved towards a more holistic thinking style. A newer branch of science called *systems science* comes to the startling realization that many of the world's manifestations are wholes supporting properties dependent on the interaction and relationships between their parts. Intriguingly, in many cases the actual make-up of the parts were not the factors found to be important in determining the behaviour of the whole. Rather, the parts of a whole were found to be somewhat interchangeable, whereas the types of interactions between the parts were the key to figuring out the whole.[8] Along this line of thinking, a flock of travelling birds and a dish of growing cells tend to assume similar patterns on account of their similar interactions and relationships, even though birds and cells are different entities all together.[9] Emergent properties were recognized and defined by scientists working from this newer perspective. The study of generalized wholes with a focus on the interactions between their parts has come to be known as *systems theory*.[10] The term *complex systems theory* is used to describe the study of wholes that are found to be composed of an immense number of parts, which of course includes the study of life's many manifestations.

A scientific *system* (indicated by the diagram of Figure 1.2) is a formalization of the way our thinking minds tend to naturally approach reality (abstracting, isolating, and simplifying). This formalized view is a neatly isolated patch of something that can often be reproduced and experimentally investigated under controlled settings using methods designed to eliminate the foibles of human perception and opinion. In its most basic description a system is comprised of: the stuff of the system; the environment, which represents everything else in the universe; and a boundary which separates this stuff from the environment. Environmental influences may act to change the so-called *state* of the system. These influences are called the *system inputs*. The system in turn may have effects on its environment. These are called the *system outputs*. Sometimes, in a process called *feedback*, the system outputs have an effect on the system inputs. Feedback is a mechanism enabling the system to regulate itself.

A system is a quintessential, formalized abstraction of reality. Systems scientists focus on defining and studying all sorts of different possibilities for the items contained within the system. For instance, the system scientist may think about how the system behaves when there are different relationships between the various items in the system, how it acts when different influences arise from the environment, or the effects of different types of feedback on the system's ultimate behaviours. Computers have enabled systems science to blossom, for they can do the hard thinking for the scientist. In systems science, rather than carrying out experiments in a laboratory (a controlled real world environment), computer simulations are used to create virtual

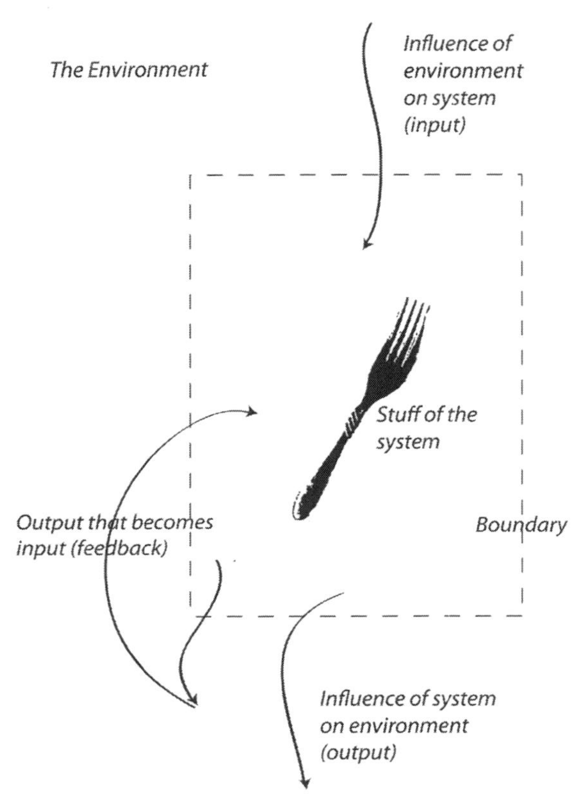


*Figure 1.2*

outcomes that are compared with what's happening in the real world situation. This means that the phenomenon can be considered in its natural context.

The study of very complex systems has become a recent inspiration for some scientists. We typically think of something that's highly ordered and consisting of only a few parts, such as a single piece of perfectly square white paper, as being 'simple' and easy to describe (Figure 1.3-A). We might think of something that has a very large number of parts as being complex. However, this isn't necessarily the case, as even for things with an immense number of parts, if they're completely disordered, it's also easy to describe them. Whether it be

*Figure 1.3*

the static on a television screen receiving no signal, or the rush of a turbulent waterfall, we call these randomized things *noise*, and scientists can describe them using only a few parameters (Figure 1.3-B). It's things that exhibit intricate patterning, such as the pattern of a leaf, that are considered truly complex and require the special attention of complex systems scientists (Figure 1.3-C). The complexity of a system is related to the length of the simplest possible description of it, and the patterns in biological organisms have intricate organizations spanning many different scales, making them very difficult to describe, and therefore, truly complex.[11]

A major goal of complex systems scientists is to account for emergent properties in terms of the parts of their supporting whole. But for complex systems consisting of so many parts, how is it possible to even approach and define the problem? How can we get an intuitive sense of how the interactions of the parts can account for the emergent properties of the whole?

Without going into too much detail, hopefully we can give you a sense of how emergent properties can be accounted for in terms of the interactions of the parts by briefly discussing a type of complex system known as a *spin system*.[12] Spin systems were designed to reflect the make-up and behaviour of intrinsically magnetic materials such as iron. However, due to the interchangeable nature of system parts, spin-systems can also be used as a general sort of tool to account for the behaviour of many other things in the world, such as ants organizing themselves to find food and even earthquake predictions.[13] A large piece of magnetic material is actually composed of an unfathomable

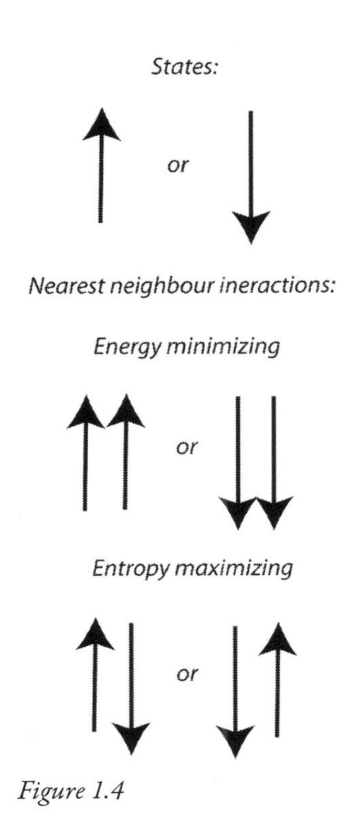

*Figure 1.4*

number of little atoms that are each very tiny (quantum) magnets called *spins*. The existence of magnetism in the large piece of material depends on the organization of all of these tiny spins. Therefore, the manifestation of magnetism in the large piece of iron is seen as an emergent property. Each of the spins can be represented as a little arrow, having a direction in which it 'points' (this is shown in Figure 1.4). The large piece of material is only magnetic when all of the spins are pointing in the same direction (as shown in Figure 1.5). If the spins are pointing every which way, then the property of magnetism in the large piece cannot manifest as it is cancelled out by the lack of organization of the spin parts (Figure 1.5).

From the whole system perspective, the piece of material either exhibits magnetism, or it exhibits no magnetism at all. These are referred to as the two phases of the magnet, and are analogous to the phases of matter (solid, liquid, gas). The phase of the whole iron bar depends on the application of an environmental influence. When the whole material is heated, it changes abruptly from magnetic to non-magnetic at a precise and consistent temperature. This is similar to water which transforms from ice to water at a precise temperature of 0°C. It is the relationships between the spins and the context supplied by the environment that determine the behaviour of the whole. Let's see how this works. Although the whole material is composed of a large number of spins, we can account for its behaviour by confining our perspective to only one of these spins and is nearest neighbours. When spins are side by side like this, they prefer to point in the same direction as it minimizes the energy of the whole system, which is desirable and comfortable for the universe. However, if all of the many spins in the large magnet are to spontaneously align, this is

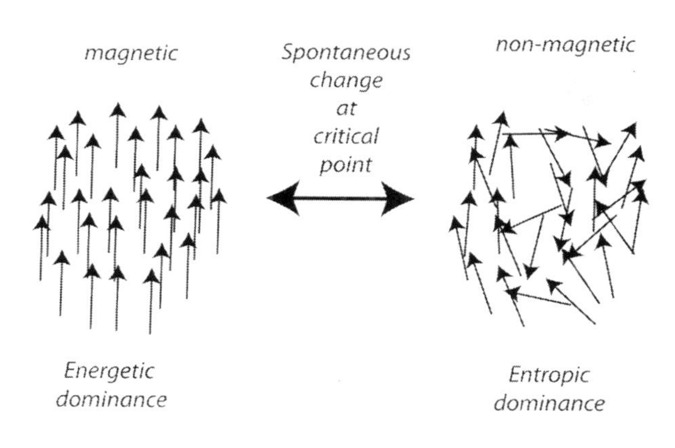


*Figure 1.5*

upsetting to the preference of the universe for disorder, rather than order, as the end result of a spontaneous process.

This preference of the universe for disorder is discussed in detail in Chapter 4. So the system of many different spins is subject to two competing influences: one for alignment to minimize energy, and one for random alignments to maximize disorder. At low temperatures, minimizing energy is most important, while at high temperatures, maximizing disorder wins out. The heat that the whole system receives from the environment is the factor which ultimately influences which of these competing influences will win. Thus, we can glean an understanding of a whole, complex system by only thinking about one of its parts, its interactions with its nearest neighbours, and the conditions supplied by the system's context.

The shift to thinking about complex systems is an important movement towards a more holistic thinking style in the sciences. Moreover, it not only recognizes emergent properties, but aims to account for emergent properties in terms of their supporting, interacting parts. However, while it represents a synthesis rather than a dissection, complex systems science still depends on a previous, extensive reductionist regime in order to know what the parts of the system are, and in order to characterize the type of relationships and interactions they are having in the whole system. In this, complex systems science still faces the very real problem of impossible computations, immense loads of data to handle, and the great degree of uncertainty about the relationships and interactions and the overall nature of tiny elements. This is not to say that complex system science is doomed to failure; rather, that empirical/holistic theories focussing on the behaviour of the properties and processes of the whole may make very positive contributions to our understanding and ability to work with complex systems.

If the three -isms perspective looks at a musical performance in terms of the position and velocity of air molecules, complex systems theories are performing the fascinating and valuable work of interfacing the larger scale properties of the musical performance (tempo, tone, harmonies) with the nature and relationships between the dynamics of air molecules. Still, there is room and utility for a theory of music that makes no reference at all to air molecules, but instead takes account of patterns and dynamics of sound itself and human experience of it. This is what we mean by a holistic theory. It is holistic theories of life and things living which are most suppressed in the life sciences.

## *The life sciences: from vitalism to vivisection*

How is your energy level today? Did this discussion drain or invigorate your energy? Have you ever met an 'energy vampire', those folks who lurk about waiting to feed upon our energy while talking us half to death? These phrases are all referring to a property associated with our bodies which is *life* itself. It is the property of *life-as-energy*.

Children have an inborn belief in this intrinsic 'life energy'.[14] It's been shown that before being taught otherwise, most children have their own innate biological theory based around the concept of life-as-energy. Similar belief systems are found in many different ancient religions and medical systems spanning otherwise diverse cultures. Life energy is the *qi* of Traditional Chinese Medicine and the *prana* of East Indian Ayurvedic systems. Navajo Native Americans call it, *nilchi'i*, and the Inuit, *sila*. The widespread concept of an *aura* typically describes a manifestation of life energy as a zone of visible light occurring in a variety of possible colours and layers surrounding human beings and other entities. Western New Age spirituality has flourished into a variety of groups, many centred about the concept of life energy. *The Reconnection*, *Bioenergetic Analysis*, and *Auric Healing* are all variations on the idea that physical healing and personal development can be achieved by working on life energy imbalances, disconnections, blocks, or deficiencies. Many practitioners of yoga, tai ch'i, chi kung, or reiki report a direct experience of life energy, describing it as a tingling sensation in their hands, a feeling of pressure, warmth or 'electrical' vibration.

Even amongst these diverse groups of people, a central belief in life-as-energy tends to generate relatively consistent explanations for the various phenomena that living things exhibit. Commonly, the body of a living thing, including each organ system, is seen to be maintained by life energy. This notion is the opposite of upwards causation, as the parts are seen to rely on something greater than themselves in order to function correctly. It is the heart that requires life energy in order to keep beating rhythmically; the brain that requires it to keep thinking with acuity. In turn, the body is perceived to be fleshy clothing that gets cast aside when the sustaining life energy departs, leaving it motionless, cold, and its form-based pattern destined for decomposition. Different interpretations believe the departed life energy to simply go out like a candle, to be transferred to a new physical form, or to transform to a different kind of energy. Life energy is seen as an inexhaustible ordering influence, directing the growth and development of acorns to oak trees, and single egg cells to mature human beings. It's seen to be extracted from the food we eat and the air we breathe. There's often the additional perception that various activities or actions can drain or replenish this energy. For instance, a good night's sleep will replenish life energy, while a trip to the emergency room of a hospital will likely leave one drained. Sicknesses and various disease states are often seen as deficiencies or imbalances in life energy. Alternative medicines such as Traditional Chinese Medicine's acupuncture and reiki therefore aim to restore the life energy deficiency or imbalance in the patient.

Initially, the Western sciences also maintained a belief system centring on life-as-energy. This belief system, formally introduced as a scientific model for living things in the seventeenth century, was called *vitalism*. Vitalism advocates the existence of a non-material based life energy (also called the vital aspect, life force, vital spark, spirit, and so on) which is fundamentally responsible for the processes of a living thing. Yet, an additional idea became closely associated with early vitalist theory, and unfortunately, the fate of vitalism as a scientific theory came to rest on proof of this idea. Let's see what this was all about.

Organic substances are those kinds of matter found only in the bodies of living things and include the material that makes up our skin, our fat, and our muscle. Inorganic substances are those found only in the world of the non-living (silver, glass, gold), or found in both the living and non-living (water, iron, table salt). In the days before a more complete understanding of how the properties of mater were related to atoms and molecules, organic substances were believed to be fundamentally different from inorganic substances on account of the unique activity of life energy. People came to this idea as they'd observed organic substances to change irreversibly with the application of heat, whereas non-organic substances went through the aforementioned phase-transitions from solid to liquid to gas. Today, the 'irreversible' behaviour of the organic materials in response to heat is what we call cooking, and occurs as organic materials are composed of additional layers of intricate structures, such as interconnected layers of cells and large many-molecule superstructures, all of which are damaged and tangled by heating. If the organic superstructures are seen like neatly rolled balls of wool, heating is to pull out and tangle up the fibers into a real big mess. In inorganic substances these superstructures are typically absent, and their molecules or atoms can undergo classical phase changes.

However, in the seventeenth century, organic substances were considered special on account of the life energy associated with them. The response of organic substances to heat was explained by imagining life energy to be irreversibly lost with heating. Life energy was seen to be the sole factor responsible for creating organic materials and making them behave differently, and therefore, it was expected that no organic materials could be synthesized in a laboratory from inorganic starting materials in the absence of life energy. While it represents only one idea of a whole vitalist theory, vitalism somehow became hinged on this notion of life energy manufacturing unique organic substances. Therefore, in 1828 when German chemist Friedrich Wöhler synthesized urea (a component of urine) from inorganic starting materials, vitalism was seen to be fatally undermined as a scientific theory, as this assumption that life energy creates unique substances was disproven. Yet, this one idea does not form the basis of the theory of life-as-energy! As we discussed in the opening paragraph of this section, if energy is that which 'makes things happen' there are numerous fundamental features of a living thing that can be seen in terms of life as an energy. That the synthesis of organic substances from inorganic starting materials is taken to be a serious blow to scientific acceptance of life-as-energy is an example of a historical circumstance which greatly influenced the subsequent path of the life sciences.

The notion of vital energy seemed to become extraneous with the discovery of the cell as the prime unit composing organ systems; the further identification, isolation and synthesis of other biological substances; and the ensuing understanding of the mechanisms behind cell function, organ function, and how organ systems interrelate to form the organism as a whole. Perhaps the last gasp of vitalist theory came in the mid-twentieth century with the discovery of molecular-based genetic material. Genes were (and often still are) believed to be akin to instruction manuals that specify all of the physical, and some behavioural, features of living things. Thus, by deciphering the information in genetic material, a complete knowledge of a living thing is anticipated.

Yet scientific resistance to the concept of life-as-energy and other holistic theories may be rooted in more than just scientific grounds. Scientific resistance may be due to social perception and political factors that have tainted the mere mention of life-as-energy and related holistic theories by associating it with implicit irrationality and thus, scientific shame. Holism describes an alternative approach to the study of a phenomenon which focusses on the patterns of the whole and no single underlying element is seen as determinative. But even the term 'holism' has become a dirty word in many scientific communities. This is possibly due to the association of holism with the discredited vitalist movement of the sixteenth to twentieth centuries, but may have more to do with the use of the term by New Age communities who are seen by scientists to embrace various claims without following the scientific testing methodology that grants them credibility. Holism and vitalism also had some very bad company, as they formed a major part of Nazi ideology during World War II.[15]

Today, the notion of life-as-energy still remains with us in the secular world, referred to in everyday language and discussions, and as a foundation concept in numerous 'alternative' medicines, therapies, spiritualities and religions. Yet the notion of life-as-energy has been entirely rejected from modern, mainstream scientific views. The scientific view of life has evolved, solidified, and remains reinforced as the cynical, clinical mainstream view, where living things are ultimately seen as fascinating chemical machines burbling out heat and motion from the molecular scale interactions occurring in the chemical soup of their bodies. The 'real' scientists snicker at those who still support the age-old concept of life-as-energy, dismissing it as naïve, foolish, outdated and inherently non-scientific. In chapters to come we will consider how a re-interpretation of the notion of life-as-energy as a scientifically respectable theory may be achievable. But before we go on to see how the gap between these two viewpoints might be bridged, let's take a better look at life and things living to re-appreciate the magnificent forms and self-creation processes lying all around and even in us.

# 2. Life and Things Living

*Beyond the rigor of the simple Euclidean regularities beloved by technologists and architects, there remains most (or all) of nature. Nature imperfectly round, never flat or square, linear only for infinitesimal distances, and stubbornly abnormal. Nature flowing, crawling, flying, weeping and in apparent disarray. Nature beyond precise measurement, and comprehensible only as sensation and system.*

~ **Bill Mollison**

A viable seed lies covered by a thin layer of warm, moist soil. Within a matter of days, its initially simple, spherical shape develops into a plant form with root, stem, leaf, and flower systems, each demonstrating an intricate patterning so precise that mathematical formulae can often describe them. In a mother's womb a fertilized egg consisting of a single cell ultimately develops into a human being comprised of about 100 trillion cells set into elaborate patterns supporting amazing properties such as emotion, thought and consciousness itself. There is no miniature plant form contained inside the seed, and no miniature human being contained within the fertilized egg. Rather, the single cells that represent the beginning stage of life forms divide, changing in form and function from their initial state. Ultimately, a great many cells find themselves in a pattern of form and function that is tens of thousands of times larger than they are. This process, formally called *morphogenesis* has not been fully explained scientifically. A scientist can still not tell you explicitly how the patterning seen in a leaf's veins develop, or how the developing cells 'know' how to form the appropriate tissue in the appropriate places to generate the length of the digits of your fingers and segments of your arm.

Life and things living exhibit many additional fascinating phenomena that remain poorly understood by science. In the Amazon rainforest, while branches of neighbouring trees may intermingle, the treetops of the canopy are distinctly separated by a few feet. How the trees can sense and respond to each other's presence remains unknown. Perhaps most perplexing are inherited migration instincts of various bird and butterfly species. Many humans believe that we are thinking creatures capable of relating complex information to one another, but that other species largely rely on inherited instinct that compels them to perform certain behaviours. Insects like butterflies are perceived to be very simple and of negligible 'intelligence' or instinctual capacity. Yet in North America, Monarch butterflies migrate some 3,000 km every year to the same over-wintering spots used for generations. What's most remarkable about this journey is that the lifespan of each generation is too short to ever make the return trip. How the naïve butterflies 'know' to complete this remarkable migration remains a mystery.

Perhaps we often forget how relatively simple a human-made object such as a light-bulb is in comparison to a living thing like an oak tree. Moreover, it seems to be largely taken for granted that we can succinctly and knowledgeably discuss the production of light from a light-bulb, while we can say relatively little about how an oak tree develops from an acorn. The success of the thinking styles of the three -isms (reductionism, mechanism and materialism) may have conditioned many of us to the comfortable notion that the various wondrous features of living things have already been, or are well on the way to being scientifically parameterized. This isn't exactly the case, and there are outstanding features of living things which seriously confound the perspective fashioned by the three -isms.

Let's embark on a whirlwind tour of life and things living, to have a look at the perplexing features of biological forms, and discuss what is known about nature's strategies for self-creation.

## *Life's creations*

*This phenomenon of growth has to do with the way crystals grow, and the daisy can be considered a living crystal.*
**~ R. Jean**

Living things are collections of individually acting units, which come together to form larger individually acting units, which come together to form still larger individually acting units, and so on and so forth. What we may agree to be non-living matter (water, proteins, fats, and minerals) comes together to form living cells; living cells come together to form tissues; living tissues come together to form an organism; living organisms come together to form societies; societies come together to form an ecosystem; and ecosystems come together to form the living planet Earth. This is quite a tapestry! Each individual cell, tissue, organism, society, ecosystem, and even the planet Earth, can be seen as a living entity exhibiting its own unique emergent properties. These individual units (cells, tissues, organisms) which are simultaneously a semi-autonomous part, and a participant in a larger collective system, are formally called *holons*.

*Hierarchical structure* describes something with numerous embedded levels of detail, and is often represented as a tree with branches, where each subsequent set of branches represents finer levels of detail (as shown in Figure 2.1). Hierarchical structure may refer to a classification system for things where more and more detail exists with progression to the finer branches, or it may refer to the actual structure of something composed of elements that are themselves made of sub-elements, perceived when the object is magnified. Living things can be seen to represent a special kind of hierarchical structure, as they are a hierarchy of holons called a *holonarchy*. We will later see that the characteristic hierarchical structures of living things can result from their self-creation mechanisms.

Living forms are often beautiful, intricate structures which exhibit what is intuitively recognizable as an ordered complexity. This ordered complexity can be quite difficult to define and describe, but it's

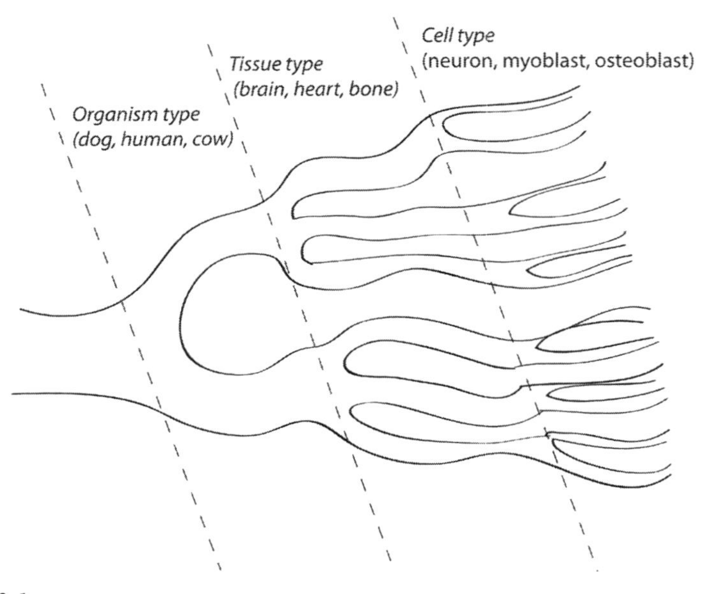


*Figure 2.1*

important as the nature of biological patterns can provide many good hints as to how they were created. Ordered complexity is in contrast to disordered complexity, which is typically studied using statistics. One way we describe the organization of a form is by its symmetries. *Symmetry* describes the ability to transform something, in one of many possible ways, only to obtain the exact same representation of the thing back again. For instance, a square has a number of symmetries. If a mirror is placed along lines passing through the middle of the square, or along the diagonal corners, the square appears in the mirror exactly as it does without the mirror, and thus, it is symmetrical along this line (as shown in the top image of Figure 2.2). If the square is rotated a specific amount, the square appears exactly the same as it did before rotating. These symmetries of the square are aptly called *mirror-plane* and *rotational* symmetries. A wave exhibits a different kind of symmetry called *translational symmetry*. Many waves are constructed by repeating the same unit over and over again. Therefore, if the wave is shifted (translated) by the length of the repeated unit, then the same form of the wave appears back again (bottom image of Figure 2.2). Until recently, scientists focussed on the above types of geometric symmetries.

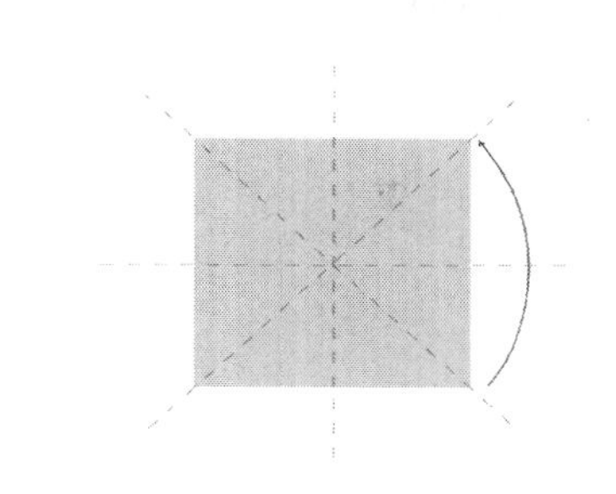

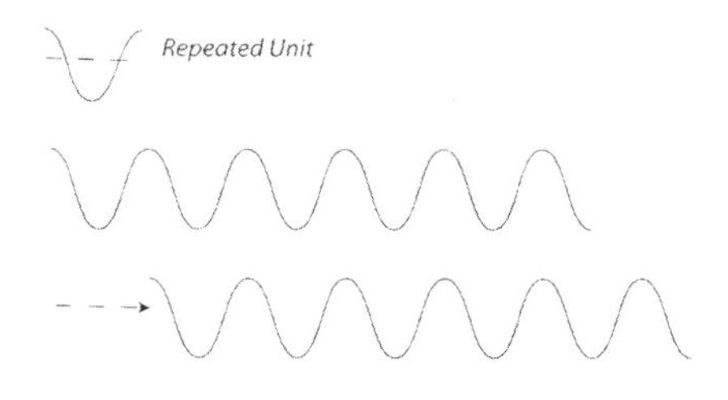


*Figure 2.2*

A special kind of hierarchical structure, and a special kind of symmetry, is exhibited by a *fractal* object. A fractal is a geometric object where each hierarchical level shares identical features with the next. Therefore, the fractal appears the same at different magnifications, which is to say at different levels of *scale.* As magnifying the fractal by a specific amount returns the same image, a fractal has *symmetry of scale.*[1] This symmetry of scale makes fractals appear exceptionally intricate to the human eye. A common example of a fractal is an object called the *Sierpinski Triangle* (Figure 2.3).

The interesting thing about fractals is that their form is intrinsically associated with the underlying mechanism that created it. For instance, one way to create a fractal is to repeat a process iteratively. The Sierpinski Triangle can be created by iteratively drawing an inverted triangle within the body of an original, larger triangle (see Figure 2.4). This process is repeated for all upright triangles that can be seen in the resulting structure. If this continues over and over and over again, endlessly, a perfectly fractal object is the result.

Many biological forms have fractal structures.[2] This is what gives them a particularly intricate and complex appearance. One of the best examples of a fractal in nature is the fern (Figure 2.5). Notice that the entire structure of the fern plant is the same in all of the individual sets-of-sets of leaves, is again similar in the individual sets of leaves, and is then mirrored in the vein pattern of the leaves themselves. While

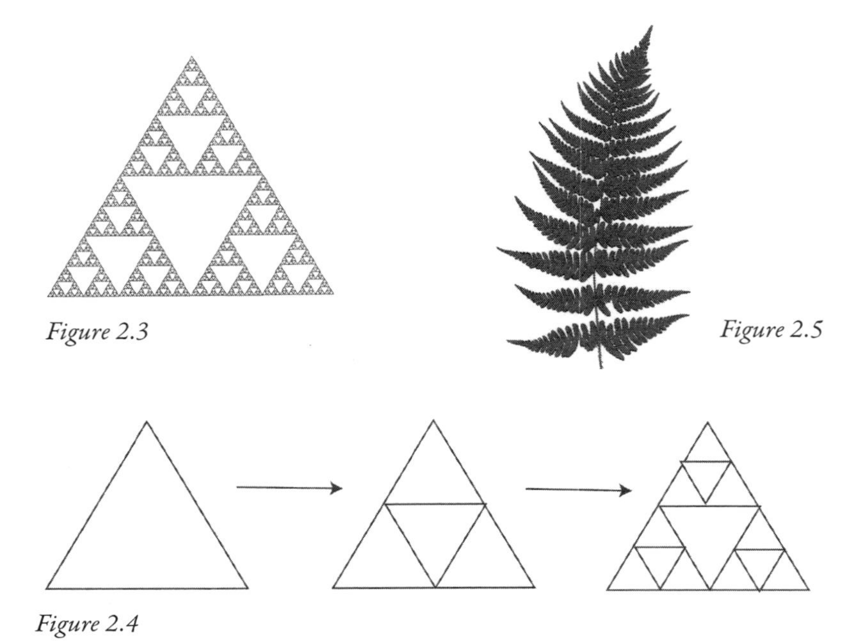

*Figure 2.3*

*Figure 2.5*

*Figure 2.4*

the form of the fern closely matches fractals created by mathematical algorithms, scientists still do not know how or why the fern assumes this particular geometry.

What is fascinating to many people is that the patterns created by living things such as the fern often possess mathematically precise features. This is very intriguing as it indicates that there are some underlying mechanisms involved in the creation of the form, rather than it being a purely random process. Moreover, this intricate precision has led some to believe that these forms of nature are sacred patterns intelligently designed by a creative deity. Various kinds of precise symmetries, spirals, helices, and fractal structures are commonly found in sea and snail shells, plant architectures, and animal forms. Plants are a particularly rich source of interesting patterns. In plants, you may have noticed that the positioning of leaves on a stem can assume quite precise organizations.[3] Plant leaves often form symmetrically facing one another on the stem (Figure 2.6-A); in helical arrangements with well defined, consistent spacing and angles (Figure 2.6-B); and in whorled patterns where leaves grow from a common point and maintain a fixed angle of separation between them (Figure 2.6-C).

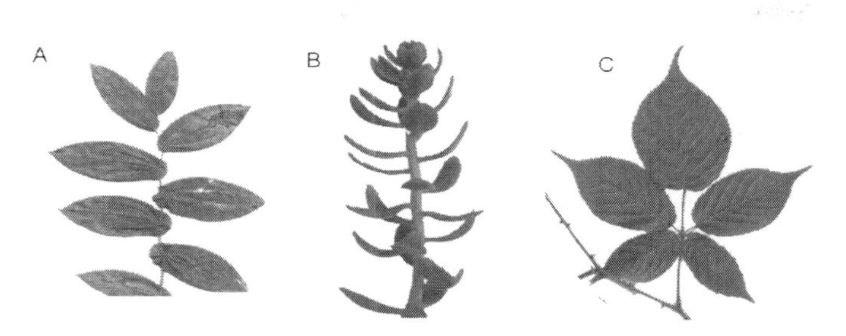


*Figure 2.6*

A seashell such as a chambered nautilus is described by a mathematical function called a *logarithmic spiral* — a form that is found repeated throughout nature in such things as diverse as a ram's horn and the form of our Milky Way galaxy (see Figure 2.7). This repetition of common archetypal forms is also evident in plants, which is exemplified by the forms of Figure 2.8.

What is even more compelling is that during development, living things are generally subject to many environmental inconsistencies. A germinating seed experiences changes in soil temperature, moisture, variable light levels and directions, and mechanical impulses and vibrations. Temperature, moisture and light are all factors required for, and of influence on, the development of the life form, and thus, one would think that variability in these factors would lead to disruption of the developmental mechanisms underway, leading to randomized structures. However, biological structures are remarkably resilient to

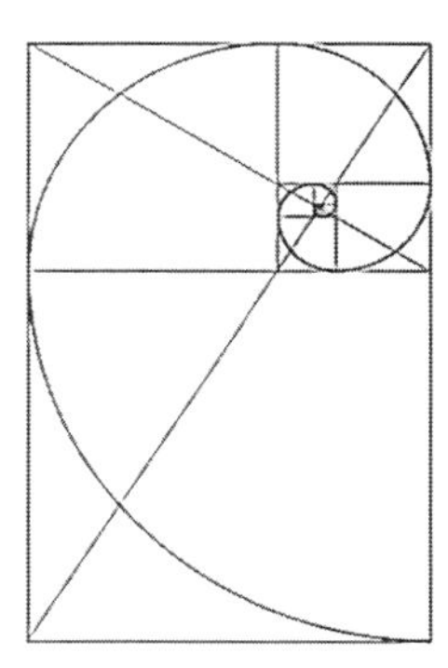

*Figure 2.7*

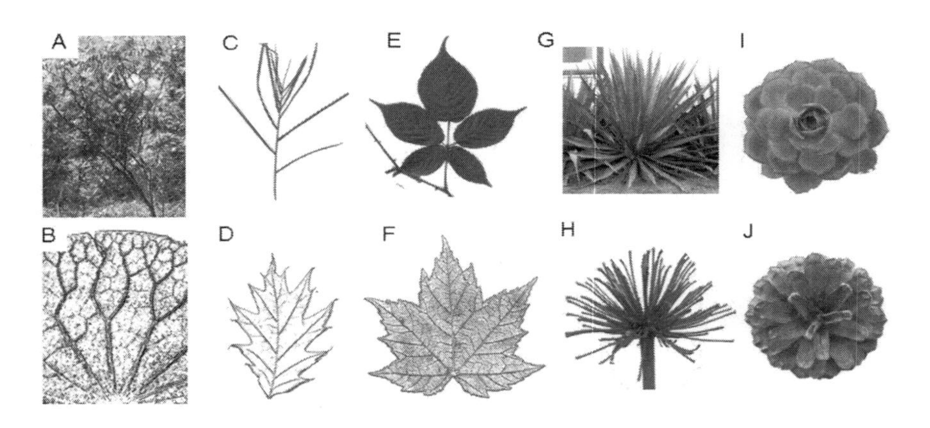


*Figure 2.8*

changes in their environment, and highly conserved patterns are most often created in spite of these random influences. Notice, for instance, that the pattern characteristic of a maple leaf is conserved in leaves of the same tree and in leaves from different trees of the same species (Figure 2.9).

How does the living thing that begins as a single cell wind up in such elaborate pattern states, especially when it's exposed to a wash of random influences from its environment? While science doesn't yet have a full answer to this question, let's examine some of the concepts from the three -isms perspective, which we do suspect to be behind the astonishing self-assembly processes of living things.

## *Life's methods*

Imagine pouring water and shining sunlight on to a pile of bricks, glass, nails, and wood, and watching them jump into exactly the right places to form an elaborate house. Of course, this is not our everyday experience with non-living materials. When we build a house, we build each level of the structure with directed energy input. We stack each brick in place with a focussed effort. And, even when the structure is complete, it must be maintained with further efforts to prevent the natural tendency of the

*Figure 2.9*

house to slowly turn into a pile of rubble. This is not at all the case for the development of a living thing. Living things self-assemble.

*Self-assembly* describes a fascinating process in which a collection of items comes together under its own impetus to form a larger, integrated, comprehensive structure. Remarkably, living forms build themselves without directed, focussed energy inputs. Supply a viable seed with soil, air, moisture, and light, and it grows into a plant without any directed efforts to assemble and place every molecule and developing cell into the final pattern. For a time an assembled living form maintains the pattern with the capacity to heal, regenerate and renew itself. This is

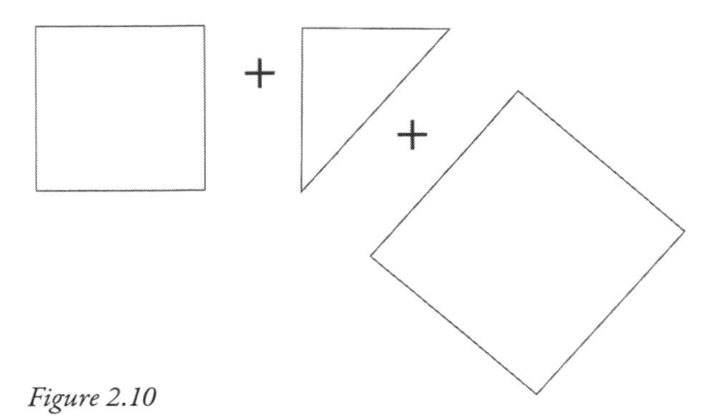

*Figure 2.10*

hardly the case for non-living materials that we commonly deal with on a day to day basis. How does nature accomplish such amazing tasks?

One basic notion behind self-assembly is the use of one level of structure as a template to organize the next level of structure, and so on and so forth. Ingeniously, 'computations' of where the next object need be put, and how big it should be, are based on constraints imposed by the existing level of structure. Thus, unlike humanity's construction methods, nature does not need to keep track of and direct the activity of each and every element in the complex process. Instead, it is the interactions between existing elements that choreograph the assembly of the whole. This iterative, bootstrapping process is why natural forms often feature hierarchical structures — each level of the structure acts as a template for the next level of the structure, eventually generating a unified whole featuring many different embedded levels of pattern.

As an example, consider a hypothetical situation where the growth of a right angled triangle is driven to occur on the face of a square, and the growth of a new square is driven to occur on the corresponding adjacent face of the triangle (as shown in Figure 2.10). Suppose that wherever there is a square, a triangle is, by some driving impetus, subject to form on the single edge of the square.[4] Correspondingly, wherever a triangle of this sort has formed, there is some driving impetus that necessitates the formation of another appropriately sized square on the adjacent face of the triangle. At each level, the existing geometric form serves to constrain the formation of the next level. The structure resulting from

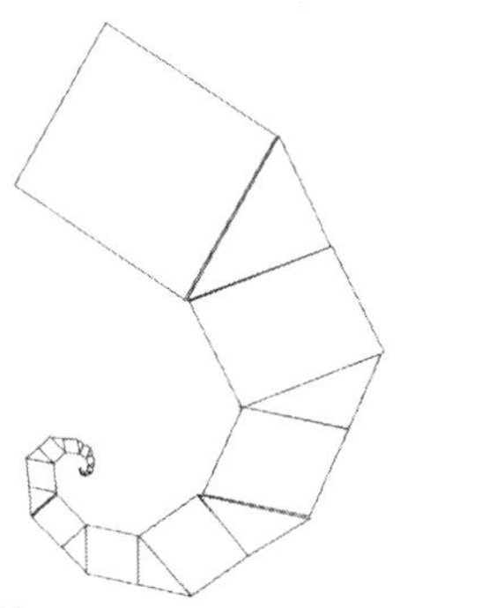

*Figure 2.11*

this iterative process (shown in Figure 2.11, left) is remarkably similar to a biological form such as the nautilus shell (Figure 2.11, right). From this we can intuit that nature can self-assemble by creating elements similar to the square and triangle, which mutually direct each other's formation into a much larger, coherent pattern.

Remarkably, it seems a similar method occurs in the development of many new life forms. In the development of an embryo, the new life form also self-generates various levels of pattern, and each level of pattern is involved in directing the development of the next, more complex level of pattern. While the processes involved are largely unknown and involve a large number of different factors that we won't discuss here, let's examine some basic things about developing organisms in a bit of detail.

Cells, like humans, respond to information in their environment, specializing in form and function to assume various roles in their tissue 'society'. Chemicals (aka molecules) are a particular kind of 'information' that cells respond to. The chemical substances that elicit such developmental changes in the cells are called *morphogens.* Cells of a particular type respond to a particular morphogen at a particular concentration to become a cell of a different type. This process of cells changing from one type to another is formally called cellular *differentiation*, and is a specialization similar to that of a young child who, exposed to information in school, may specialize to fulfil particular roles in society (a diagram of the process is shown in Figure 2.12). The

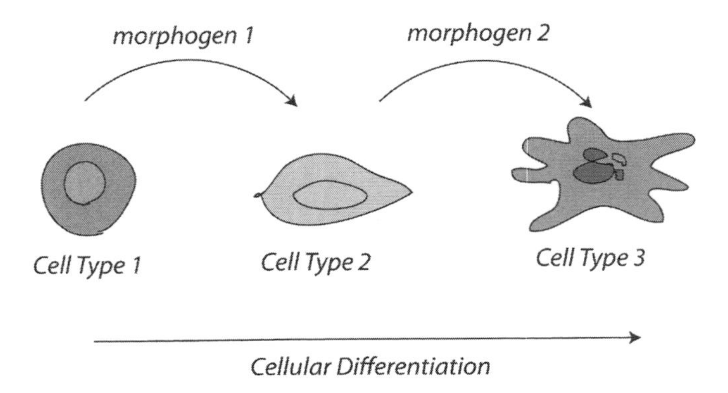


*Figure 2.12*

differentiation state refers to what kind of cell it is. For instance, red blood cells and neurons represent different differentiation states for the cells of a body, but both originated from a common precursor cell somewhere along their developmental paths.

Independent of what kind of cell it is in the body, each cell contains a complete set of *genetic information*. In its essence, genetic information is a sequence of molecules, where each molecule is analogous to a letter in a language (there are 4 unique molecular letters in the genetic language). The molecular 'letters' spell out 'words' that tell the cell how to go about making a substance called a *protein*. Proteins make up much of the structural stuff of a living thing (skin, hair, muscles, bones, and so on), but also act as messengers for the cells, and facilitate various chemical reactions in the cell. The genetic information is partitioned into chemical phrases called *genes*. Each gene tells the cell to prepare a specific kind of protein (there are nearly an infinite number of different possible proteins). There are many thousands of unique genes in the genetic code of an organism, but each cell in the living thing contains the same, complete set of genetic information (aside from the sperm and egg cells which contain half the genetic material of other cells).

The basic role of genetic information is to produce proteins, but not all genes are active at any particular point in the cell's lifecycle, so only a few proteins are being produced at any point in time. The kinds of products that the cell is producing help to define which kind of cell it is (that is, they specify the cell's differentiation state). The products manufactured by the cell depend on which parts of the cell's genetic code are active at any point in time. It turns out that morphogens can be a special kind of protein produced by cells, which acts to switch portions of a cell's genetic code on or off when a cell is exposed to it. Therefore, a morphogen, which cells produce themselves, changes the proteins that the cell will be producing, and thereby influences the cell's differentiation state. When cells specialize to a new differentiation state, they can produce different morphogens, which leads to further change in the differentiation state.

The developmental biologist Hans Meinhardt has done a lot of work with the idea that cells in a developing organism have the ability to create and maintain a pattern of morphogen within the cluster. Since cells exposed to morphogen will change to a new differentiation state, the pattern of morphogen created in the cell cluster generates a corresponding pattern of differentiated cells (this is illustrated in Figure 2.13). This *chemical patterning* is one of a number of important ways that a life form is known to self-organize.[5]

Yet, when a many-celled life form begins it starts out very simply, with no miniature starter-pattern embedded inside the egg or seed, so just how does any patterning get started at all? The cell will generally divide to form a small clump of cells that are all pretty much the same in shape, size, and unspecialized differentiation state (Figure 2.14). There appears to be nothing special or particularly interesting about

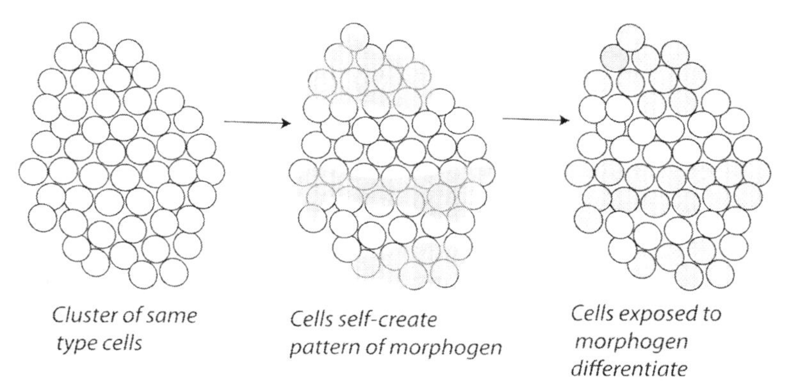


*Figure 2.13*

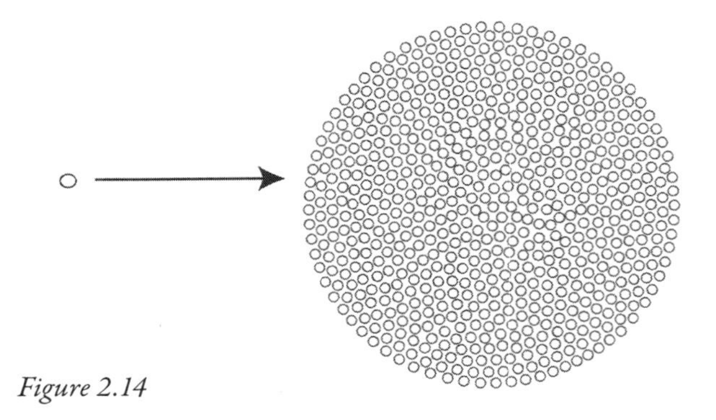

*Figure 2.14*

this clump of cells, in fact, it all starts off looking pretty boring, without any hint of the intricate patterning that's to come. One picture for what happens next was postulated by Meinhardt, who has shown that all of a sudden there can be some change in the environment surrounding the clump. Perhaps the temperature changes a bit in one place and not the other, or it becomes a little more acidic at one particular spot. With this tiny random signal a single cell of this clump suddenly starts behaving differently than all the rest. This is signified by a new gene being turned on in only the cell that experienced the environmental perturbation. This special cell begins producing the first morphogen and that morphogen spreads throughout to influence all other cells in the cluster and stabilize a pattern of morphogens in the once boring clump. The world has suddenly changed, completely and irreversibly, by that one cell changing, for now there are entirely new possibilities available for the clump. It's analogous to how human reality changed after the discovery of the wheel. Suddenly, entirely new things like carts, pulleys, gears, clocks, bicycles, and cars, all things that used wheels, became possible, where before, they had been unimaginable. It's the same thing for the clump of cells after the first changes, for the symmetry of the spherical clump is forever broken. A pole, consisting of a small cluster of differentiated cells, can now form at the spot on the clump where the first special cell began its work. This pole can work with the new geometry to create even more sophisticated structures such as a central axis and an opposing pole. Once two poles and a central axis have formed, even more sophisticated mechanisms can be generated (Figure 2.15).

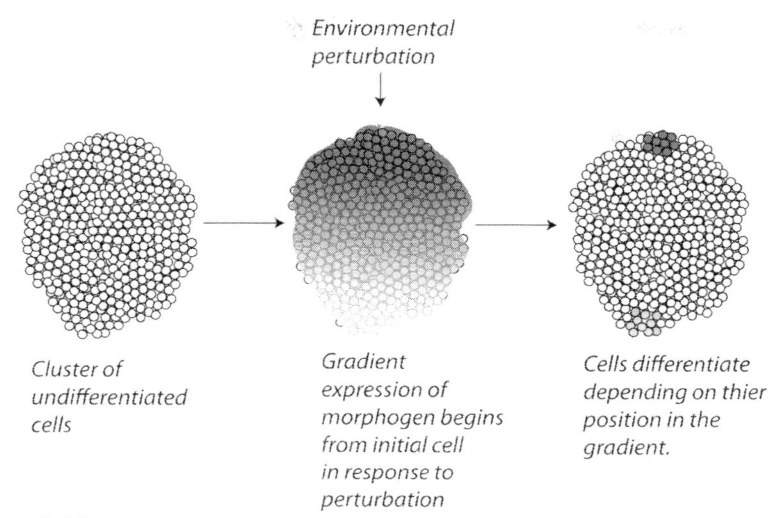


*Figure 2.15*

Unfortunately, the above cascade of change quickly becomes very complicated, as it involves the expression of many genes, the production of many morphogens, and the production of many factors involved in pattern making. All the while cells continue to divide, the cluster continues to grow, and even cascades of programmed cell death are initiated. In addition, there are not only chemical morphogens involved in changing differentiation state, but a number of other factors such as mechanical stress, chemicals that attract the direction cells move in, and stages of programmed cell death which come into play.[6] The situation quickly becomes too complex for effective study from the ground up, even with today's immense computational-power. Thus, the details of morphogenesis are far from being sorted, and represent an immense research effort in the modern biological sciences.

## *Precision from randomness: the chaos game*

It may be a little difficult to wrap one's mind about the strange 'bootstrapping' processes that underlie a living form's mysterious self-assembly processes. A little something called the *chaos game* can help give us an intuitive feel for how the intricate, reproducible structures of

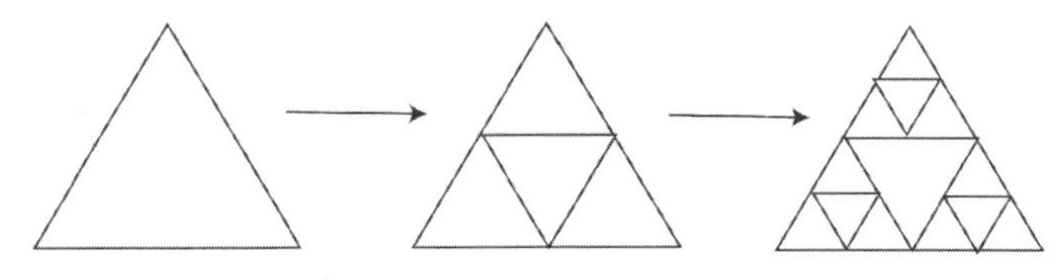

*Figure 2.16*

biological form can arise as emergent patterns, even in a situation that appears to be a wash of random influences.

In mathematics, an *algorithm* is a set of instructions, much like a cooking recipe, that will produce some structure or outcome if carried out faithfully. It's interesting that a few entirely different algorithms can produce the same fractal object, such as the Sierpinski triangle. Some of these methods are called: the linear creation method, the multiple reduction copy method, and the chaos game.

We've already been introduced to the linear creation method of generating a fractal (Figure 2.16). The algorithm behind this method is described by the steps:

(1) Draw a triangle;
(2) Draw an inverted triangle within all upright triangles in the structure;
(3) Repeat step (2), an infinite number of times ...

The multiple reduction copy method is another less intuitive, but perhaps more interesting way to make a fractal (Figure 2.17). For this case, one can begin with any initial image. Three copies of this image are made, which are each reduced to half of their original size. These copies are placed at the vertices of a triangle. This process is repeated with the newly obtained image, placing it at the same vertices as the original triangle. After many iterations of this procedure, the result is once again, the Sierpinski Triangle.[7]

Another way to create a fractal is by playing the chaos game (Figure 2.18). The chaos game is the slowest, least intuitive way to construct a fractal, but is probably the most fascinating and illustrative of the notion that precise, intricate structures can emerge from a seemingly

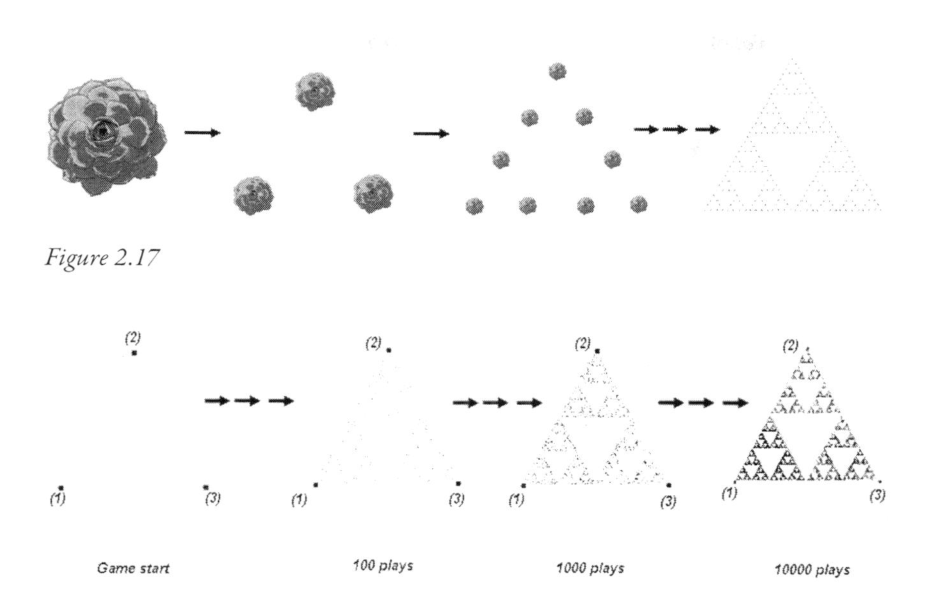


*Figure 2.17*

*Figure 2.18*

random course of events.[8] To play the chaos game, begin with three nodes at the vertices of a triangle, labelled (1), (2) and (3). These three nodes remain fixed throughout the entire game. Now, randomly select a point anywhere on the page, within or even outside of the area that would define a triangle if nodes (1)-(2)-(3) were connected. Imagine you have a spinning wheel with three sections and an indicator on it. Spinning this wheel you can obtain a number from (1) to (3). The only rule of the chaos game is that for whatever number you get when you spin the wheel, you'll place a point halfway between the number of the node that corresponds to the number you obtained, and the point made in the previous round of the game. So for instance, if you obtain a '2', you will draw a point halfway between node (2) and the initial point. In the next round if you get a '1' when spinning the wheel, you will place a point halfway between the node (1) and the point that was just made in the previous round. If you continue to play this game for a long, long time (thousands of rounds) then the Sierpinski Triangle will again emerge as a result!

A first point of examining these three algorithms is to show how a relatively simple set of instructions can, when repeated many times, generate

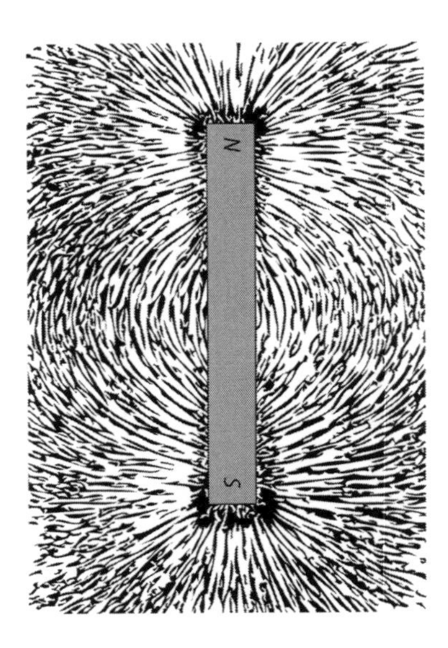


*Figure 2.19*

a precise and intricate structure with long range features. While patterns in nature do not look like a Sierpinski Triangle, and are not likely to follow the same algorithms described here in any fashion, the creation of the Sierpinski triangle from these simple instruction sets gives us a feel for how patterns in nature can be created by rules and constraints established by the developing organism and its environment.

Another thing to learn from this is the idea of the final structure (Sierpinski Triangle) as an *attractor pattern*.[9] Although we're using three very different kinds of instruction sets, we start with clear open spaces, and eventually end up with the same Sierpinski structure for all of the different algorithms. For the case of the multiple reduction copy machine, and especially the randomized chaos game, it even appears as if the system is drawn in towards, or is 'attracted' to the Sierpinski structure. This 'attraction' to the final pattern is conceptually similar to the orientation of iron filings in the magnetic field about a bar magnet (Figure 2.19). The generation of the Sierpinski structure is also similar to the process of biological development, where intricate, long range forms manifest out of a homogeneous space, and may also seem to be 'attracted' by some kind of overriding 'force field'. In this simplistic analogy we could consider that the points in the manifesting Sierpinski triangle represent cells in a growing collective. Similarly, the initial growing clump of cells is 'attracted' to the form of maple leaf.

The chaos game algorithm also points out the importance of context. The placement of new elements in the form depends on the state of the existing form. New elements form in the developing *context* supplied by the existing system. Intuitively, we can see that if we studied the chaos game focussing with a reductionist view on the parts (points) on their own, it would be impossible to explain how the final pattern is formed.

A final thing to take home from this example is a better intuitive grasp of what emergent properties are, and how little is scientifically known about them. In this case, the Sierpinski triangle represents some biological form that's manifested from a developmental process. This biological form goes on to support emergent properties and processes that are entirely new to the functional, assembled collective (life, perception, and thought). Scientists are currently working on an understanding of the algorithms behind the manifestation of the biological form (again, in this example biological form is represented by the Sierpinski Triangle), but still have a long way to go to understand the formation of these forms. The emergent properties of life arise from, and are supported by the final structure. Scientists can't even describe the development of the form yet, and thus, have no real idea yet about how these additional emergent properties are generated. Therefore, in a developing organism, there's no definition of when the life of the organism officially begins, as there's no actual definition or measure of life as a property.

## *Genes: the book of life?*

There's a prevailing mythology in modern day life sciences, and even in secular society, that our genetic material is a 'book of life' analogous to an instruction manual or software program that specifies all of the properties of a living thing's form, and many aspects of its behaviour. Modern science has therefore jumped full stream into decoding the information in genes, and uses the gene as a basis for the life sciences, much as the atom became the basis for the physical and chemical sciences. This is generally assumed to make perfect sense given various observations and is evidenced by various technological feats. After all, genes are the inherited molecular units, the only material items to be passed on, and we can clearly observe discrete trait passage from mother and father to their offspring. We can even map out the statistical chances that a parent's children will exhibit a certain eye or hair colour, or a genetic condition such as sickle-cell anemia. Our genetic information, if left behind in shed skin or hair cells can be used to identify us and prove our presence at the scene of a crime, while a baby's genetic information can prove whether or not we are its mother or father. The genes of a plant crop can be modified to make it produce its own herbicides, become drought resistant, and to produce larger, sweeter or more storable fruit. Gene therapies promise to relieve conditions such as cystic fibrosis and muscular dystrophy. Our genes can be blamed for our cancer, obesity, depression, and even alcoholism, or can they? These observations have led to the optimistic assumption that by knowing the genes of an organism, we will know everything that there is to know about the form of the organism, what diseases it may develop, what instincts it may exhibit, and what makes it different from other life forms. The mythology that's developed around genes makes them seem to be deterministic information manuals capable of relating all of the secrets of life, if only we know how to decipher the genetic language. However, if we examine the situation a bit further, we can see how it is another mindset, and that while the gene myth has been a route to functional understanding, it too has limitations.

The surprising thing about the past few years of gene-o-centric science is that the genetic promises have, to an extent, let us down.[10] First of all, if a life form's genes are a book of life, responsible for specifying their physical and many behavioural traits, then differences between life forms, especially between complex life forms like human beings and simple life forms like single-celled bacteria, should correspond to particularly outstanding differences in the genetic information. The human genome project, taking thirteen years and costing roughly $10 US billion, was an international scientific research endeavour that determined the entire genetic code of a human being. Other related projects have determined the genetic codes of mice, fruit flies, rice plants, and *E. coli* bacteria. It was expected that human beings, being so obviously more complex in structure and function than rice, would show orders of magnitude greater number of genes than these other simpler creatures. The hard fact was that while it was anticipated that human beings have at least 100,000 genes, the final report tallied up only around 24,000 genes. A bit disturbingly, this is essentially the same number of genes that mice have. Even more disturbingly, a fruit fly was found to have about 13,000 genes. The single celled bacterium *E. coli* has 4,400 genes. Perhaps most disturbing of all is the notion that rice plants have more genes than humans, tallying in somewhere between 35,000–65,000.[11] In addition to an alarmingly low number of genes in comparison to other much simpler creatures, a surprising similarity was also observed in the kinds of genes that were exhibited. Fruit flies share 60 % of genes with humans, mice and humans are very genetically similar, and 99.99% of genetic material is the same between humans of different races. The question of why organisms are different from one another could not be answered by knowing all of the information contained within their genetic material.

Knowing the genes of an organism is similar to having a list of the isolated words that have been used in a book. Clearly, we don't know what the book is about by just knowing the different words that have been used. In life's language, the 'ordering' of these words doesn't happen by continuing to work on the level of the genetic information; rather, it involves knowing the entire context and mechanism in which cells, their genes, and the products they produce, are functioning. Genes don't specify information like a software program or instruction manual does. Genes code for protein. They participate in a context greater than themselves. For cases where a single protein is involved in manifesting a single factor, knowledge of genes appears very powerful. This may be the case for the colour of hair, eyes, or flower-petals where a single gene codes for a single pigment. When the gene is passed on from parent to child, and is functioning properly, that pigment shows up in the hair, eye, or flower petal. However, these cases are much lower in number than we initially thought, and only one small subset of the actual reality of a living thing. This can be illustrated by considering the number of new medical drugs that were expected to be developed knowing the complete human genome. In reality, very few new drugs have been developed since the completion of the human genome project, as the number of cases involving single, simple sets of genes and corresponding cell products is very limited.[12]

A prime example of the importance of context was given by Brian Goodwin in his book *How the Leopard Changed Its Spots*.[13] When water flows out the bathtub drain, the direction that the vortex spins in depends on the existing movements of the bath water that may be caused by your shifting leg, the way you stand, or a breeze blowing in through the bathroom window. To prove this to yourself, you can experiment and note that the direction of the water's spin can actually be changed by swirling the water flow in an opposite way. The direction that the draining water spins in doesn't depend on what the liquid in the tub is made out of; it depends on the movements of the liquid occurring in a greater context. Thus, if a bathtub were filled with a liquid, and I told you what the molecular parts of that liquid were, you would not be able to tell me the direction that the draining water vortex would spin in from that information alone. The three -isms have a particular focus which tends to forget about this need to look for and define a new context in which to apprehend the phenomena of study. The development of new context requires an allowance for imagination, innovation and a released insistence on the three -isms mandate.

Knowing genes hasn't helped in understanding the differences between organisms. It hasn't helped to identify many medicines and therapies, which were highly anticipated with the completion of the human genome project. There remains no understanding of how instinct can be inherited from genes. Nor has an understanding of the functioning of a single neuron yet assisted in understanding consciousness, awareness, or love. This occurs for the same reason that we can't predict the direction of the bathtub water's vortex knowing the composition of the liquid alone. Keep in mind that the intention here is not to refute or deny the functional understandings and significant achievements that have come from gene-based thinking, but rather, to point out its limitations and to imply that there is ample room for innovation and discovery in the science of life.

## *Physics: imagination in action*

*Physics is imagination in a straightjacket.*
**~ John Moffat**

Physics and chemistry, the so called 'hard' sciences, are perhaps the most trusted of all sciences, as they seem to uncover the 'laws' of the universe. These branches of science are often seen as being primarily founded on reductionalist approaches, where the fundamental material unit is the atom, and much effort exists to determine the properties of the parts of the part of the atom (electrons, protons, quarks), in order to derive even greater knowledge of fundamental universal laws. The life sciences followed in what was assumed to be the reductionalist, materialist, mechanistic footsteps of their granddaddy hard sciences. It is expected by some that life and things living will eventually be described and understood in terms of the fundamental laws of physics and chemistry. All in all, it seems as though the life sciences have ossified into a material literalism where only theories stated in terms of genes and molecules are deemed valid.

However, a reductionist monopoly involving a material literalism is not necessarily the case for other branches of science. A closer examination of the concepts of physics for instance reveals that reductionism does not dominate; rather, as we'll explore in the next chapter, a holism that employs metaphoric mappings and works with wholes and imagination is well tolerated and serves an essential role in contextualizing axiomatic concepts such as energy, and to work effectively with phenomena such as heat and magnetism at scale relevant to everyday human activities. This holistic thinking style holds the vitality and imagination that are so desperately required to bring the life back to the life sciences.

# 3. A Swim in the Ocean of Thoughts

*Science does not know its debit to imagination.*
**~ Ralph Waldo Emerson**

In past chapters, we've noted that some human beings interpret the world from two vantage points. It seems we are simultaneously ghosts in the formless internal ocean of beliefs and feelings called our subjective experience, and participants in an external or objective reality of material forms and events. Scientists tend to see themselves in their subjective oceans attempting to obtain an accurate picture of the one objective reality. From this platform, scientific knowledge takes on an air of truth, fact, and transcendence from the human condition. Emotions and beliefs based on emotions are perceived as contaminants of a pure assessment of the one objective reality. Perched at this particular perspective, the modern scientific mindset also uses a conceptual pair of glasses based on three thinking styles called reductionism, mechanism and materialism. These three -isms appear to dominate the scientific view of everything, from galaxies to grapefruit.

We've seen that technologies based on the three -isms often catastrophically fail when functioning in the bigger picture represented by an ecosystem or organism's whole body. This indicates that the thinking styles have trouble considering the larger context of interconnected wholes. We've also seen that there are aspects of living things that confound the existing scientific mindset of the three -isms. Of particular note are emergent properties (life, thought, emotion), which simultaneously arise from the activity of a collection of underlying elements, but may also dominate the behaviour of the same set of elements (cyclic causality). To effectively work with emergent properties (like music for example), we've noted that the development of new *context* may be very helpful. We've also noted a strong focus on genetic information as an instruction manual specifying physical features, biological processes, and intuitive behaviours. Yet, knowing genes hasn't really helped to clarify why there are such outstanding differences between organisms, nor has it helped to develop a plethora of new medicines, nor new understandings of life and consciousness. This is because context is again so important. Knowing the genes of an organism is similar to knowing a list of all of the isolated words used in a particular book, in absence of any sentence structure. Just as this list of words will not tell you much about what the book is about, neither will a list of genes tell us what a living thing is all about. In life's 'language', the 'ordering' of these words doesn't happen by continuing to work on the level of the genetic information; rather, it involves knowing the activities of cells, their genes, and the function of gene products acting in the context of the larger collective of cells developing and acting in space and time (a systems science view). On account of the many poorly understood features of living things, and the limitations of the gene-o-centric viewpoint, it's apparent that there's ample room for innovation and discovery in the life sciences. It's time for the paradigm of the ghost in the ocean and the lens of the three -isms to be supplemented with new thinking styles, but what are they?

Perhaps the very basic notion of subjective experience and objective truth is a very long-established mindset. It seems subjective and objective experiences may actually be quite inextricably connected to one another, and though one may try very hard, it's rather difficult, if not impossible, to draw a clean line between the two. Thoughts, originating in subjective experience, seem to be generated from our interaction with the external, objective reality. Paradoxically, thoughts structure the way we perceive and approach reality, and to some extent, this determines what we will experience in objective reality. The consequence of this paradox is that our concepts, even our scientifically derived concepts, are not absolute truths transcendental of the human condition. We will consider that the formation of new concepts very often involves imagination — an *imaginative rationality* — though we may not be fully aware of it yet. So while the three -isms appear to dominate the way scientists think about the universe, if we look a little closer there are many ideas, some particularly fundamental ideas, which have no reductionist/materialist basis, and can be seen to have developed using processes associated with imaginative rationality.

In this chapter we'll consider the remarkable possibility that some of our concepts spontaneously unfold and develop in ways that are not unlike those of self-assembling life forms previously discussed. We'll look at imaginative rationality in terms of *conceptual metaphors,* which are inherent features of the human mind that structure a new and unusual set of experiences (the unknown) using what we have previously and commonly experienced (the known) as a model.[1] Thus, the known is used as a *template* for the unknown. This is similar to the templating process used as a self-assembling strategy of developing life forms, as described in Chapter 2. Just as a life form tends to develop a hierarchical structure as a result of templating, we will see that our conceptual systems also develop a hierarchical structuring. We'll consider how this conceptual templating of the unknown to the known is already used scientifically and is an alternative thinking style to the three -isms. Accepting this use of conceptual metaphor shakes the paradigm of the ghost in the ocean, as it anchors even scientific concepts in embodied human experience (the most known thing of all). Most importantly, we will see that the recognition of an ideal conceptual metaphor is the cooking-up of fresh *context* in which something can be studied by proven means. Finally, we re-examine some of the intuitive, but outdated, rejected, and ignored concepts of the life sciences in this new light, seeing them as befitting conceptual metaphors for various aspects of life and things living. In seeing these older concepts as conceptual metaphors we can consider them to be just as acceptable as other foundation concepts in physics, and in subsequent chapters we will go on to see if they can be developed to bring forth some much needed new understandings of life and things living.

## *Conceptual metaphors*

*Metaphor is not merely a matter of language. It is a matter of conceptual structure. And conceptual structure is not merely a matter of the intellect — it involves all the natural dimensions of our experience, including aspects of our sense experiences: colour, shape, texture, sound, etc.*

**~ George Lackoff & Mark Johnson (from *Metaphors We Live By*)**

Thoughts are formless structures, but just like structures that have form, thoughts can seem solid and distinct, or fluid and flowing. Thoughts can be fresh and juicy as summer cherries, stale as old white bread, or as stimulating as a madras curry. Thoughts can take us somewhere, get us somewhere, or lead us around in circles. Thoughts are tools to build something new, or to form something to shelter and protect us. We can cook thoughts up. We can build thought-things with basic thoughts, putting them together into more complex structures that are our concepts, theories and beliefs. Thoughts are born and thoughts develop and thoughts grow old and die. We can dress thoughts up with jargon and big words to make them look sophisticated and professional. We can clear space in our head for our thoughts. We can have cluttered and muddled minds because of too many thoughts.

Notice that we *know* our thoughts and speak of them as we know and speak of objects such as food, tools, buildings and people. We are understanding thoughts as *objects,* which to be meaningful, must point to and match up with our experiences. We understand symbols, words and phrases as *containers* for thoughts. Communication is the act of putting thought 'objects' into symbolic 'containers' and sending them along to a recipient who knows how to open the box and get at the meaning that is contained inside. Understanding is *seeing.* Awareness is the *light* that allows us to see. Consciousness is the inner *space* in which this all happens.

This is all the way that it is because some of our fundamental concepts are made by drawing the more known (familiar or concrete) to the unknown (unusual or abstract). Often this understanding of the unknown in terms of the known has been done without consciously recognizing the underlying picture or model we have used to structure our experiences. This connecting of the known to the unknown is formally called a *conceptual metaphor* (see Figure 3.1). Of course, the actual nature of thought, symbol, understanding, awareness, and consciousness are not really that of our everyday world structures, containers, visual perception, light, and outer spaces, yet the mind employs conceptual metaphor to draw these more tangible, perceptible elements of our experience to the more abstract. For the time being, it seems this is the only way that we can know of our thoughts, symbols, understandings and consciousness.

A more familiar type of metaphor is the so-called *poetic metaphor*, found in phrases such as: '*the wind whispered softly to her*', '*Father Time had been unkind to him*'. Poetic metaphors are best thought of as conceptual cosmetics, akin to a paint job on a house. Like a paint job they might make the concept they're associated with stand out with clarity, look better, be more emotionally compelling, or be more interesting. However, it's easy to separate the poetic metaphor from the underly-

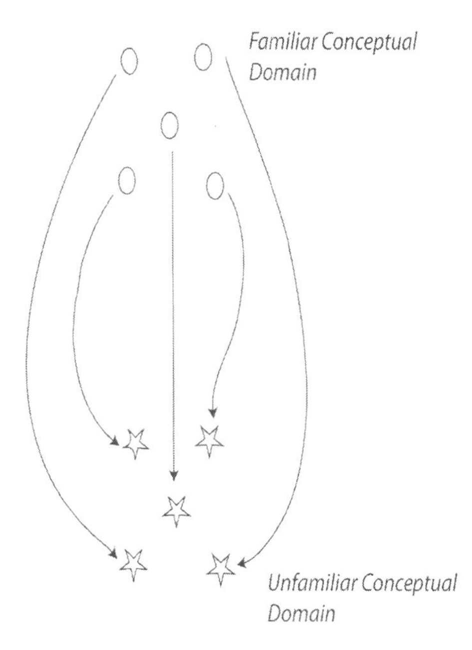


*Figure 3.1*

ing concept it's associated with. In contrast, conceptual metaphors are more like the wooden framework on which the entire form and structure of the house exists. The conceptual metaphor and the concept become identical, in the sense that the features of the underlying model of the conceptual metaphor are used to *structure* the new idea. This means that when a conceptual metaphor is used to structure a concept, features of the underlying modeling picture become systematically referred to in all of the language used to talk about the new idea. We can see this happening in the first paragraph of this section, where a conceptual metaphor structures thoughts as objects. Throughout the passage thoughts are systematically seen as various kinds of material object, and similar things that can be done with objects, can be done with thoughts. Unlike poetic metaphors, when a conceptual metaphor is used to structure our experiences, our minds actually make the two situations conceptually equivalent, and we are therefore rarely aware that a comparison is being made at all. Conceptual metaphors allow us to interpret all aspects of our existence. This extends from ideas of so-called objective reality (science) to more obvious manifestations of the human condition such as love.

Conceptual metaphors can be identified by examining the language used to speak about things. The experience of romantic love for instance, can be conceptually structured using common models such as.

Love is a living being
Love is magic
Love is a physical force [2]

In each case, the metaphoric model structuring our particular experience of love shows up in the language we use when we think or talk about love. If we are using the metaphor that love is a living being, then we assess whether the relationship is *healthy*, if our love is *dying*, or if it is *growing* stronger and more vital. Health, death, growth and vitality are all concepts associated with living things. If love is magic, then we are caught mysteriously and helplessly under the power of the *spell* of love. If we see love as a physical force then the language revolves about terms used in the sciences and we may say that there's a tremendous *attraction* between us; that we have good *energy* together; that we're *bound* by love; and that we can feel the *electricity* when we kiss.

Another experience well structured by conceptual metaphors is that of time.[3] We commonly view time from the metaphoric framework that it is a moving thing. Time is *rushing* past. Time *got away* from me. The time is *moving slowly* today. These phrases all imply that time is an actual substance that moves. The whole system of language is consistent with this conceptual picture or model for time. Another common structuring of time is as a limited resource and valued commodity. When we structure time using this picture, we say that this is *wasting* my time; that the blender will *save* you hours in the kitchen. We ask how you are *spending* your time. We say the man with a brain tumour is living off *borrowed* time. Again, the entire system of language reflects the equivalency of time and money. Using this conceptual metaphor we can do the same things with time that we do with money: we can waste it, save it, spend it and borrow it. These pictures for time are *meaningful* because our human experience and cultural environment supply a context that is consistent with the use of models of time as a moving substance or valued commodity.

To be meaningful, metaphoric models must be consistent with the context supplied by experience (the set of observations pertaining to the phenomenon). Apart from this condition, there is no intrinsic 'goodness' or 'badness' to the particular model we use to structure our experiences. The metaphors may in fact be interchangeable (such is the case for time as fluid versus time as commodity) depending on how the concept needs to be used. However, it is important to recognize that the model used to structure experience will greatly influence how we perceive, and consequentially act in, the world, as it in turn establishes a context in which to perceive and act. For instance, conceptual metaphors for our lives illustrate how our perceptions and actions are structured and influenced by metaphoric models. Perhaps some people may view life as journey, others as a gamble or game, while others with growth and development. Someone who sees life as a journey may be driven to always be getting somewhere, to be on the right track, to know where they're going in life, and may feel troubled by circumstances that result in feeling lost or tired along the way. Someone who structures life as a game may take chances in order to win big, make bets, and rely on risk and chance to manifest opportunity. Someone who structures life as an agricultural enterprise will perhaps be more content to stay in one place, to plant seeds of opportunity, and take time to let things develop. Each of these structuring pictures supplies a different context that will lead to different values, opinions, pursuits and approaches to a human lifetime.

## *Conceptual metaphors in science: cooking up new context*

*New metaphors are capable of creating new understanding and therefore new realities.*
**~ George Lackoff & Mark Johnson (from *Metaphors We Live By*)**

We've previously discussed the essence of science as a methodical approach to reality that maps out various consistencies of the universe by decontaminating our observations of the world from our beliefs and feelings about it. We see for instance that particular phenomena, such as a ball rolling downhill, are attributed to something we now call the force of gravity. Gravity is a property of the universe recognized by science. It's possible to determine the effects of gravity by observing quantifiable, reproducible and predictable events associated with it. It is inconsequential whether or not one believes that gravity exists, or if one wishes that gravity did not exist. The effects of gravity remain the same predictable, reproducible effects no matter who you are, or what you may believe. It therefore seems that gravity, and all scientific laws that are based around the concept, are something that's independent, or transcendental of, human experience. From this there's the expectation that an advanced alien species from a far away galaxy would have the same mathematics and understanding of gravity. What we have come to call gravity does seem real, to belong to an objective reality, but the argument presented here is that the way we come to understand gravity (and other things) is inexorably embedded in an inherently *human* perspective. This is not an attempt to blow the notion of objectivity right our of the water, just to suggest that there's a bit more involved in the development of functional human concepts. In many cases it can be quite clearly seen that our basic scientific concepts, even our tried and true scientific laws, facts and axioms, rest on a platform of human experience as embodied minds existing on planet Earth; not as minds accessing a singular, transcendental, objective reality.[4]

Let's begin by noticing a particular slipperiness to some concepts. Even those notions we might consider to be very basic, essential facts can slide away from the face of reality when considered in a different context to the one they were invented to be used in. Take the concept of *distance* or *length* as an example. The distance spanned between two points seems to be a fundamental, useful and frequently used measure. How big is it? How long is it? How far did we travel? We get out our ruler and have our answer. We've travelled from A to B and it was so many units of length (Figure 3.2). There is a sense of absoluteness, truth and fact in the idea of length measurements.

Einstein, in his work on special relativity, showed that if we were to ride in a spaceship travelling near the speed of light, we may have some troubles with our length measurements. Most of us don't have the opportunity to travel in a spaceship moving that fast, most likely because such a thing doesn't currently exist, and so we don't run into this kind of trouble on a day-to-day basis. Length remains a fairly solid idea and we may feel pretty confident to say that length is a 'real world thing', not some loose conceptual model that we're using to get by.

Yet travel in a fast spaceship isn't the only way that our conventional idea of length breaks down. Consider measuring the distance between A to B on a very *wavy* line with a ruler that is very big, another one that is smaller, and a final one that is very small and allows you to get into all of the nooks and crannies of the line (as shown in Figure 3.3). Changing the size of a ruler as such is known as changing the *resolution* of a measuring instrument. As the resolution of the ruler gets smaller, the length measurement between A and B

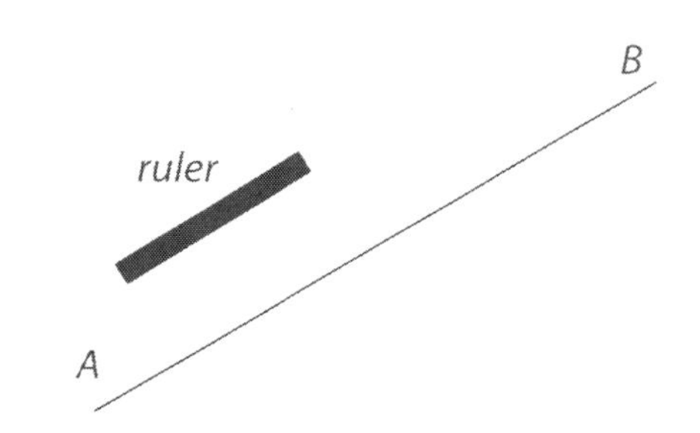


*Figure 3.2*

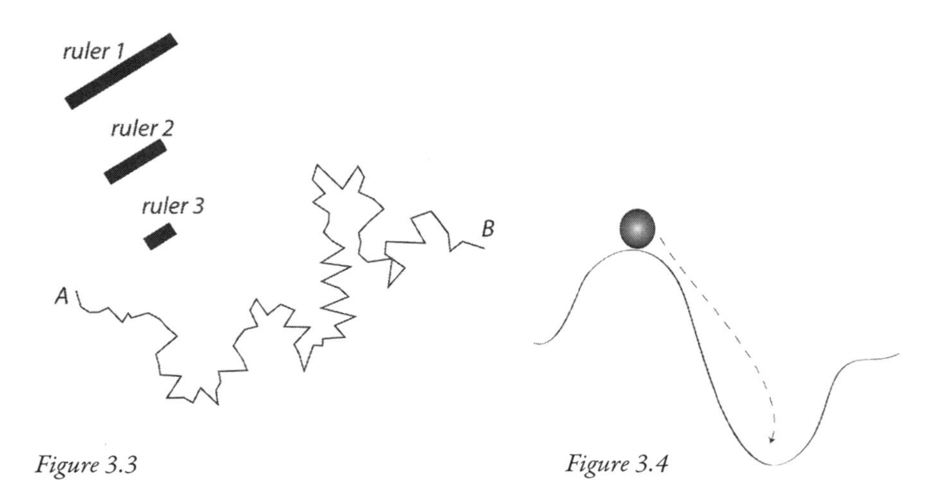


*Figure 3.3* *Figure 3.4*

gets significantly larger. What if magnifying the line segment between A and B showed still higher levels of detail? How long is the line now? How far is the distance between A and B? What do we even mean by length, if there is an infinite amount of detail to the line and an ultimately high resolution ruler to measure it? The solid concept of an absolute length or distance falls apart. The concept of length or distance can of course still be used for most things in the day-to-day activities of most humans. When the concept of length or distance was initially invented it was used to relate how much cloth was desired, or how long it would take to travel from one village to the next. The concept of length is based on the worldly context of human experience. In the past, the measure of the surface area of a piece of small intestine, the roughness of a piece of metal, or the precise distance along a rocky, craggy shoreline were not deemed useful; however, in today's world, these quantities are sometimes desired, and so our concept of length has been augmented to include these more unusual situations.

So concepts, even scientific ones, can on occasion look a bit slippery, and a reason for this might be because scientific thinking makes rich use of conceptual metaphor. What we're going to see is that some of the most basic scientific concepts have been obtained by imaginatively drawing upon experience — primordial, embodied, and uniquely human experience — to obtain models that effectively structure a new concept. This use of conceptual metaphor in science is an intrinsically creative, non-reductionist, and holistic approach to obtaining a functional understanding of a phenomenon. Moreover, it engineers a whole new context in which to work with a phenomenon.

Let's see how conceptual metaphors are used in the making of scientific concepts. As a first example, let's consider how human experiences with gravity on the surface of planet Earth are used to structure observations of two interacting magnets. Gravity is an everyday experience for human beings as embodied entities living on planet Earth, and the effects of gravity on objects are common observations. It is for instance common experience that when a ball is placed at the top of a hill, only a small disturbance is required to initiate movement of the ball down to the lowest point of the landscape (Figure 3.4). If however the ball is placed in a valley, it will remain there unless some strong external influence acts on it to drive it out. The specific features of the landscape (for example, how steep the hill is, how deep the valley is) determine the stability of the ball in its placement, how it moves on the landscape, and how much energy must be transferred to it for it to escape from the valley. Let's refer to this common situation as the ball-on-a-hilly-landscape model.

Magnetic objects are arguably a less common experience than that of the Earth's gravity. Every magnet has two poles corresponding to the endpoints of the object, which are often referred to as the North (N) and South (S) poles. The inherent nature of these endpoints is that poles of the same character repel one another, while poles of opposite character attract one another. If two magnets, fixed at their centre point but free to rotate, are placed in a position where poles of the same character face one another (Figure 3.5 top), the slightest perturbation will cause the magnets to rotate into a configuration where poles of opposite character face one another (Figure 3.5 bottom). Once in a configuration where poles of opposite character face one another, an external force is required to twist the magnets into a new configuration due to their attraction for one another.

A functional understanding of the behaviour of these bar magnets is obtained by contextualizing observations of the magnetic objects using the ball-on-a-hilly-landscape as the underlying model in a con-

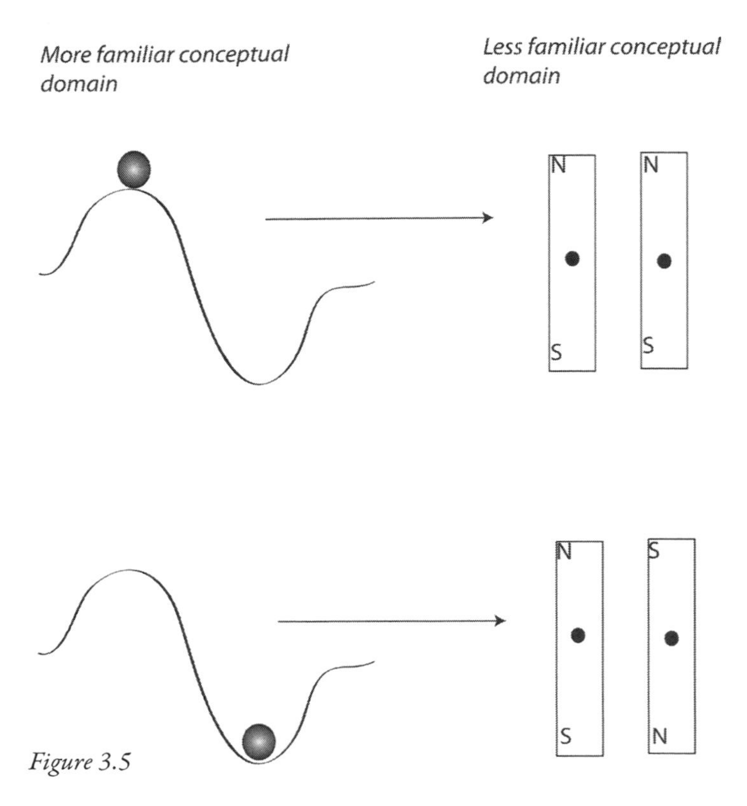


*Figure 3.5*

ceptual metaphor. Magnets in a parallel position with poles of the same character facing one another are made equivalent to the unstable situation of the ball on top of the hilly landscape (Figure 3.5 top). Magnets in a position where opposite poles face one another are made conceptually equivalent to the stable situation of the ball at the bottom of the hill (Figure 3.5 bottom). The characteristic features of the ball on a hilly landscape have been abstracted and used to structure observations of these interacting magnetic objects. The conceptual metaphor is meaningful as the observations of both situations correspond with one another within the supplied context.

This ball-on-a-hilly-landscape model is now routinely used as a conceptual metaphor to contextualize, understand, and work with many diverse situations. For instance, two ions (atoms or molecules with an electric charge) are attracted to one another, and are often understood as an unstable state directly analogous to a ball at the peak (Figure 3.6

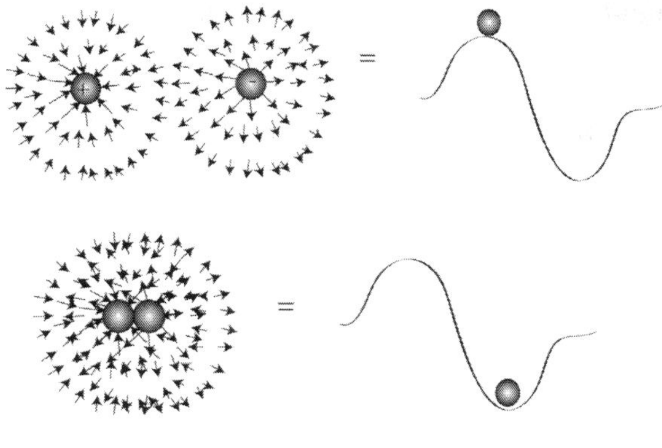

*Figure 3.6*

top). Likewise, having formed a chemical bond between one another, the ions are seen as a stable state analogous to the ball in a well (Figure 3.6 bottom).

This conceptual metaphor of the potential energy landscape is made even more sophisticated and used to structure features associated with chemical reactions, where atoms are binding or unbinding to form various clusters of atoms called molecules. When atoms form a chemical bond between one another and thus become a molecule, they have minimized their potential energy and are in a stable state analogous to a ball in a potential energy well. It may however be the case that this particular configuration of bonded atoms represents only a *local minimum* in the landscape where a deeper well, representing perhaps a new bond between different elements, exists (see Figure 3.7). If enough *activation energy* (the amount of energy required to break the existing bonds) is supplied to the system, the new configuration can be assumed, and a certain amount of active energy will be released in the process (as the second state is at a lower energy 'valley' than the first). Activation energy is analogous to a kick which thrusts the ball out of one valley, over a peak, and into another valley (see Figure 3.7). Existing atoms may now adopt the new configuration which represents a favourable decrease in their total energy overall. Thus, the spontaneous and heat-absorbing or heat-releasing nature of chemical reactions is also contextualized by further developing this ball-on-a-hilly-landscape model to include concepts such as activation energy; local and global minima (valleys); and energy release or absorption with movement to new stable states, which depends if the 'ball' is moving from a higher valley to a lower one, or the other way around (Figure 3.7).

Of course in all of these cases, it should be clear that we're no longer talking about a ball on a physical landscape acting under the force of gravity; rather, the ball-on-a-hilly-landscape is an effective conceptual metaphor to model the interactions and phenomena of the chemical system. The specific behaviour of the magnets, charged atoms, or chemically reacting atoms/molecules is related directly to the features of the 'landscape' (depth of 'valley', height of 'peak', steepness of 'walls', and so on) using a mathematical formalism. However, aside from conceptual congruities, there is no implicit connection between the situation of a ball on a hilly landscape, and the behaviour of magnets, charged objects, or other systems. In essence, a meaningful conceptual metaphor has been constructed, and a sense of understanding results from the supplied context. Note that systems have not been broken into parts in order to explain the functioning of the whole. It's also important to note that science does not recognize these situations as conceptual metaphors. Rather, what we've described as the 'ball-on-a-hilly-landscape' model is considered to be a real-world thing called *potential energy*.[5] To get a further handle on how conceptual metaphors are used in science; to compare the sense of understanding obtained using a reductionist approach, versus the holistic, imaginative employment of conceptual metaphor; and to see how intrinsically irrec-

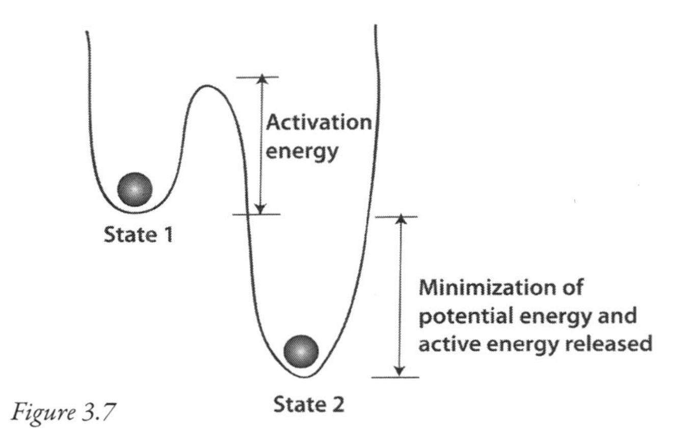


*Figure 3.7*

oncilable conceptual perspectives are acceptably used to understand the same phenomenon, let's consider the property we call *heat*. Heat is of course what makes something feel warm and fiery. An absence of heat makes something feel cold, icy or frozen.

Viewed from a holistic, metaphoric perspective, heat is a property: an ineffable, fluid-like substance (similar to time) that flows from hot to cold objects. The amount and rate of heat flow can be measured quantitatively using a device called a thermometer, which yields the quantity called temperature, usually via the relationship between volume changes of a substance (such as mercury) with heating. A comprehensive mathematics has been developed to account for heat flow, and it even has a form similar to the mathematics developed to work with the diffusion of actual material substances. For all intents and purposes at the macroscopic level of the everyday world that we humans live in, heat does appear very much like a fluidic entity flowing from hot to cold objects.

There is a completely different interpretation of heat and temperature that comes from the thinking style of the three -isms. In this approach to understanding, a piece of matter (like a rock for instance) is broken down into basic elements, which so happen to be atoms (or molecules, clusters of bound atoms). These basic parts can possess *kinetic energy*, which is the energy of motion. The more kinetic energy an entity has, the faster it moves. Looking at the rock from the perspective that it is a whole composed of many tiny parts, the energy possessed by the whole can be *coherent* or *incoherent*, denotations which refer to how all of the parts are behaving. For instance, if the rock is kicked into the air and so supplied with kinetic energy, the net motion of all of the molecules of the rock is *coherent* in the direction in which the rock is travelling (Figure 3.8, left). If however, the rock is left out in the sun, some of the sun's radiant energy may be absorbed by the rock's molecules, making them all jiggle and wiggle with more kinetic energy, but in more or less random directions. Since in this latter case the molecules move in random directions with respect to one another, the motion of the molecules is said to be *incoherent* (Figure 3.8, right). The sun has increased the incoherent kinetic energy of the rock, and this is what we perceive to be *heat*. Thus, from a reductionist perspective, heat is

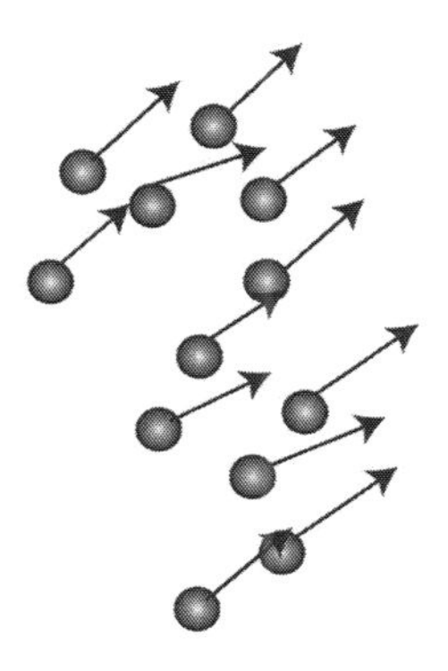

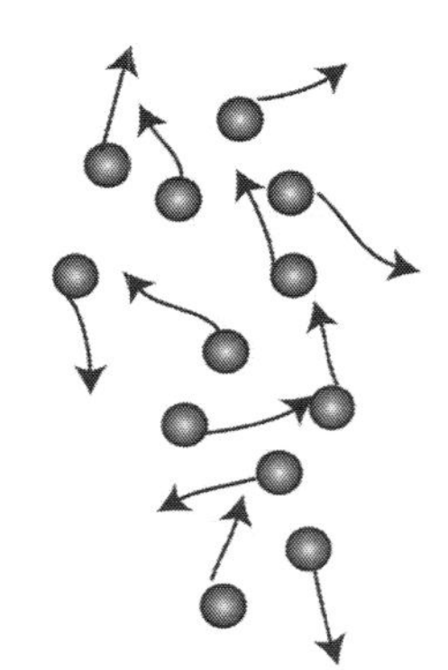

*Figure 3.8*

related to the amount of incoherent movement of the primary elements of a piece of matter. The reductionist definition of temperature is as a statistically defined quantity relating the probability of finding these basic elements possessing a certain amount of kinetic energy. At cold temperatures elements are concentrated in the states of low kinetic energy (low movement), whereas hot temperatures are characterized by a diffuse spread of elements through all possible kinetic energy states (a more even spread from low through to high movements). This reductionist perspective of heat and temperature is covered more comprehensibly in Chapter 4, pages 135–137.

The important thing to notice is that physics didn't completely reject the 'holistic' interpretation of heat-as-a-property with the development of a three -isms styled perspective in terms of the motion of the parts. Both the reductionist and holistic perceptions of heat are scientifically acceptable and mutually tolerated. Each perspective is valuable and has applicability, depending on the context in which one is working from. A reductionist view of heat and temperature is necessary when designing a semiconductor device for a computer, whereas the holistic, metaphoric view is immensely more practical when designing a cooling system for a car.

## *A tree of knowledge*

In this long example, we're going to see how scientific concepts develop like an organic thing, unfolding in full self-assembling glory like sprouting, growing, living things. We'll see a few layers of abstraction here. At the most fundamental level, a new concept is imaginatively rendered by drawing from basic human experience as an embodied mind living on the surface of planet Earth. This is what makes science dependent on human experience. At the more abstracted end of this example, we'll see how scientific concepts themselves are used as a template model for the construction of new, even more abstract, concepts via conceptual metaphor. This is an important observation as in later chapters we will be doing just this: using existing scientific concepts to structure observations of life and things living in order to develop new functional understandings of them and new context to work with them still further. For those unfamiliar with scientific and mathematical concepts, take heed that the following material may be difficult to understand. Since this section is not essential for understanding the remaining material of the book, it can easily be skipped if it frustrates you.

Cyclic and wave phenomena are an important feature of our experience. There are the cycles of day and night, seasons, phases of the moon, orbits of the planets, and the revolutions of car wheels. There are water waves, sound waves, radio waves, microwaves, heat waves, and light waves. There are heartbeats, a woman's menstrual cycle, the budding-growing-falling-and budding of deciduous tree leaves, the diurnal rhythm of a water lily opening and closing, and predator-prey population cycles.

How do we formally comprehend cyclic and wave phenomena? What formal mathematical systems of thought have we designed to work with these kinds of things? In spite of the ubiquitous nature of these phenomena, one of the most primary experiences of the human body is motion in more or less a straight line. As we walk or ride horses, bicycles or cars, our fundamental experiences with motion are as passengers moving through a box-like space with a two-dimensional plane below us and open space surrounding us. This is what is experientially most familiar, most known to us. In other words, we don't know first hand what the physical experience of being a wave is. Perhaps not surprisingly then, our observations of cyclic and wave phenomena are structured using these more familiar experiences and associated concepts of linear motion. Let's see how this works.

Our basic experiences with motion in space have been represented by something called the *Cartesian co-ordinate system* (Figure 3.9), another integral scientific legacy from René Descartes. The Cartesian co-ordinate system defines three essential directions that exist at right angles to one another. Since these form the platform in which any object in the space can be described, they are called the basis dimensions of space, commonly referred to as the 'x', 'y' and 'z' directions. The x and y directions form a flat surface called a plane (the x-y plane), which could be thought of as the surface of the Earth we traditionally walk upon. The z-direction usually represents the space above and below the plane. We can look down from the z-direction to the x-y plane and see a bird's eye view of the plane. The Cartesian co-ordinate system is a model for space as human beings commonly experience it. As such, it isn't a conceptual metaphor, but a functional abstraction of human experience into a symbolic thought-system in which some aspects of reality can be represented.

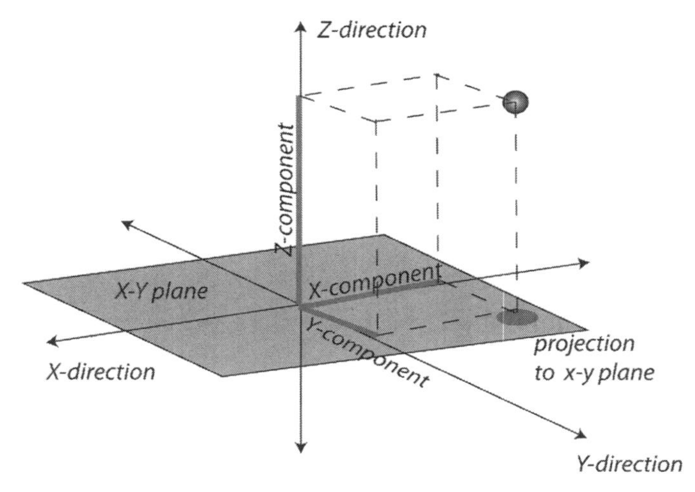


*Figure 3.9*

The three basis dimensions (that is the x, y and z directions) are the fundamental, essential ingredients which come together in different quantities to specify the location of any point in the space. It is possible (and useful) to decompose the position of any element in the space into varying amounts of each of these basis dimensions. By being able to decompose any point in space into various amounts of the three basis dimensions, we are able to tell anyone what point we are referring to by quoting the relative amount of the three basis dimensions that make up the point. The decomposition of a point into any one of its basis dimensions is analogous to the act of projecting a shadow. We can imagine that if a light is shone along one of the basis dimensions, the object in the space will cast a shadow onto the plane or axis of the complementary basis dimension (see Figure 3.9). The length between the origin of the Cartesian space and the shadow of the point is the amount of the basis dimension making up the point's position. Mathematically we have constructed tools to represent the process of making a shadow, and we call these shadows *projections*. As the point moves in space along a trajectory, the relative length of each component (projection) will change.

How is cyclic movement handled in this Cartesian space? For simplicity, let's imagine that a particle is confined to movement in the x-y plane, and look down upon the space from above (this is shown in Figure 3.10). Let's suppose that the particle is moving in a circle. In this situation, if a light is shone along the x-direction, a shadow from the particle forms along the y-axis which represents the y-projection. Similarly, if a light is shone along the y-axis, a shadow appears on the x-axis due to the presence of the particle, which represents the x-projection. If we keep track of the length of these projections as the particle moves around the circle, we would discover that the length of the projection as a function of the position of the particle traces out a wave (see Figure 3.11). The wave shape that forms when the length of the y-projection is plotted is called a *sine* wave. The wave formed by keeping track of the length of the x-projection is called a *cosine* wave. We'll refer to these as circle-based waves.

We now can see some of the layers of this particular concept tree. Wave phenomena are described in terms of a formalism based on the

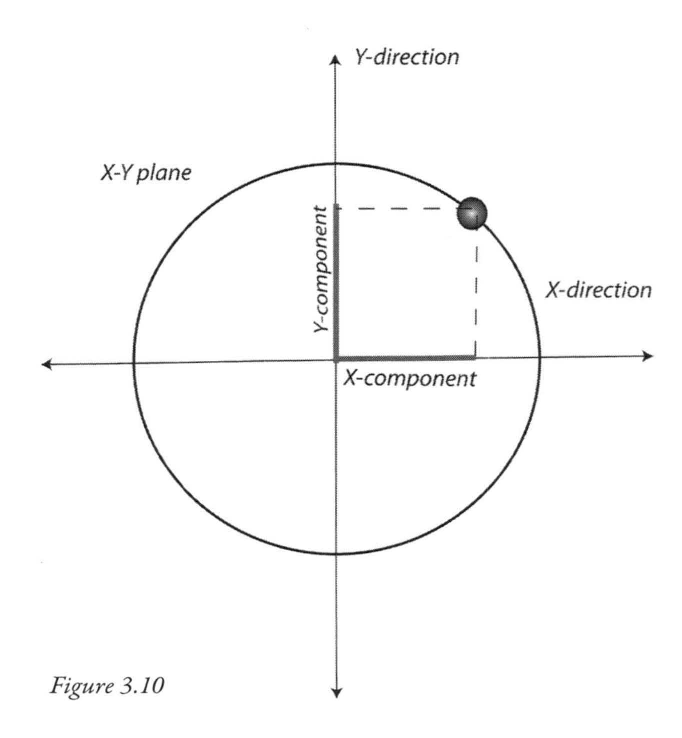


*Figure 3.10*

geometry of a circle, which in turn, is linked to the basis dimensions of Cartesian space, which are, in turn, grounded in our basic physical experiences as embodied human forms moving in space.

We can make things more interesting by introducing time into the model. If we count how often the point goes around the circle in a unit of time, we come up with a new concept called *frequency*. As the motion of the point of the circle is directly related to the form of the wave, the frequency of a wave represents the number of repeating cycles which occur in a unit of time (Figure 3.12). The wave's amplitude describes the height of the wave's crests and troughs.

Frequency is a very important feature of wave phenomena. For example, many of our sensory perceptions respond to the frequency of wave phenomena occurring in the environment. Sound, for instance, is a wave occurring in a space of air molecules. The frequency of the sound wave is perceived as a different sound pitch, where high frequency waves correspond with high pitches, and low frequency

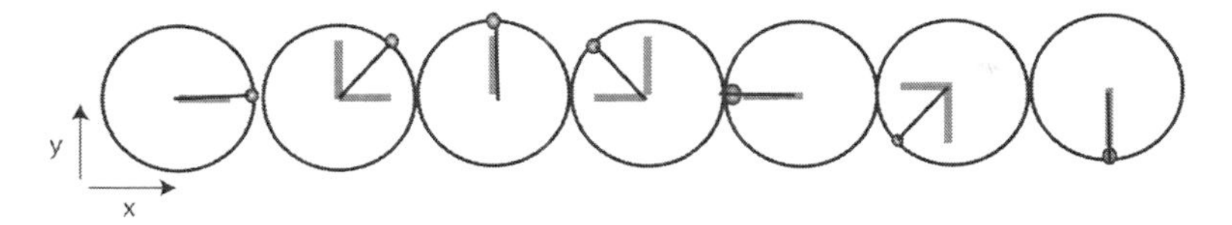


*Cosine Wave:*

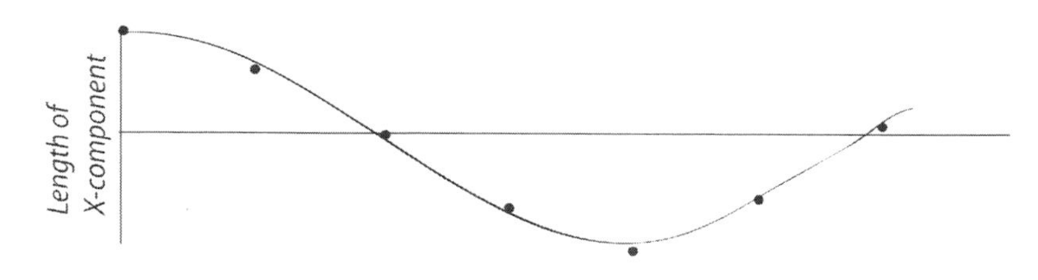


*Sine Wave:*

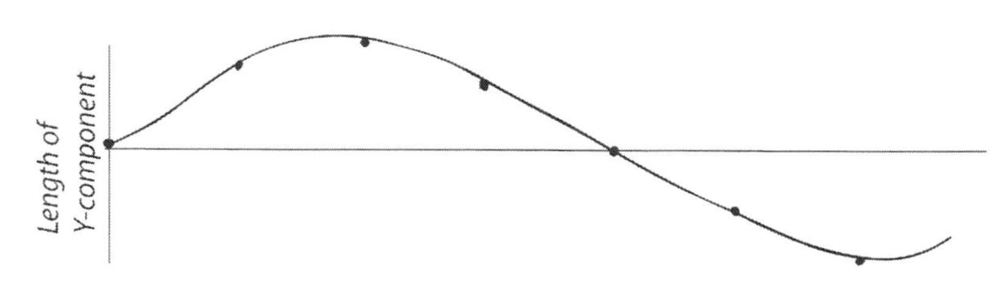


*Figure 3.11*

waves with low pitched sound. Light is another wave phenomenon. The different colours that we see correspond to different frequencies of the vibrating light wave, where purple is a high frequency wave and red is a low frequency wave. A concept directly related to frequency is that of *wavelength*, which describes the length of the wave's repeating cycle (see Figure 3.12). Frequency/wavelength can help us to account for some important properties of a wave based phenomenon, and therefore, it is a feature that is worth knowing a lot about, especially, how to work with it most effectively.

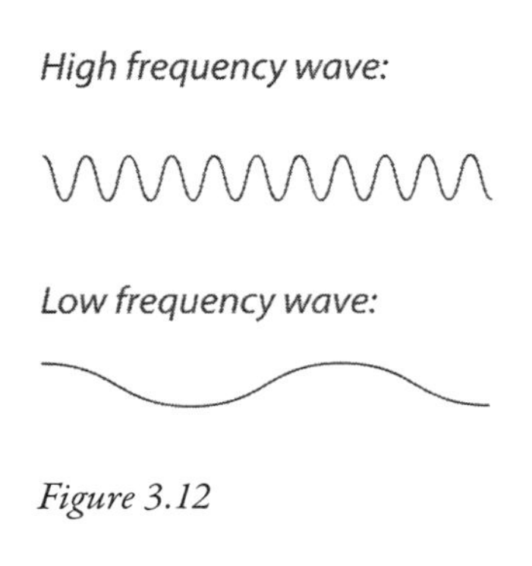


*Figure 3.12*

It was observed that any general wave phenomenon could be described as a combination of the circle-based sine and cosine waves, the form of which we've just introduced. Sine and cosine waves of different frequencies add up in different amounts to create any shape of wave at all (Figure 3.13). Correspondingly, any wave-like signal can be broken down and described as a combination of circle-based waves.

It's at this point that our knowledge tree develops a fractal (self-similar) character, as a conceptual metaphor is used to structure and facilitate working with wave phenomena. Interestingly, the conceptual metaphor uses the original notion of Cartesian space to facilitate work with wave phenomena! The basis dimensions which define *physical space* can be used as a model in a conceptual metaphor structuring observations of wave phenomena. Quite literally, a physical space comprised of fundamental basis dimensions is extended to establish the concept of a *frequency space* comprised of fundamental, essential basis waves (see Figure 3.14). As any wave can be seen as made up of circle-based waves of different frequencies, a circle-based wave at a particular frequency becomes analogous to a basis dimension of Cartesian space. Analogous to the case of physical space, any type of wave can be represented as a 'point' in frequency space and can similarly be decomposed into the basis circular-waves at a particular frequency, again by casting shadows (making projections). In this metaphor, it is the height (amplitude) of the wave at a particular frequency which corresponds to the length of the frequency component along a particular axis.

In Cartesian space, only three basis dimensions are required to define the space. For the case of frequency space, any frequency of circle-based wave may be possible; therefore, frequency space can have an infinite number of dimensions. While it is impossible for us to visualize anything with more than three dimensions, it is still possible to use the idea of three dimensional space, and all of the activities that go on within it such as translation, combination, and projection, as a model

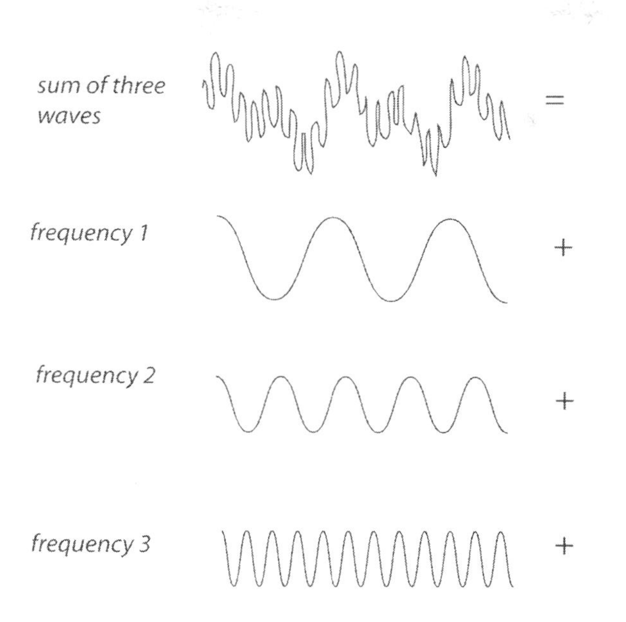


*Figure 3.13*

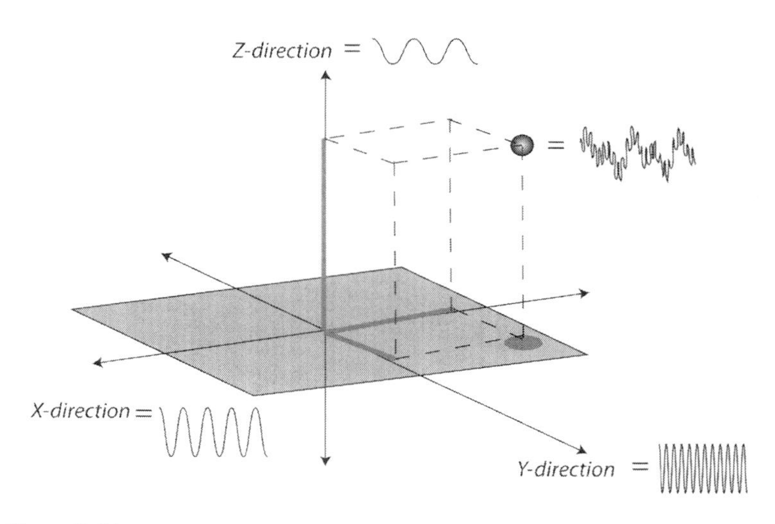


*Figure 3.14*

to structure experiences with waves. This allows us to understand and work more effectively with wave combinations at different frequencies and the important phenomena (such as light and sound) associated with them.

Now, what we ultimately have is a set of concepts allowing us to work with various things happening in space, from the representation of objects, to complex wave phenomena. This set of concepts has a multi-layered structure with some self-similarity between the layers, which has developed as conceptual layers can act as a template for the next, more abstract and sophisticated layer (a diagram of the example outlined here is shown in Figure 3.15). Our fundamental experiences as embodied minds on planet Earth are first used to structure basic concepts (the Cartesian co-ordinate system). These perceptually grounded concepts are then used to structure more abstract observations (motion of points in space, definition of waves in terms of points moving on circle). These more abstract concepts are used to structure even more abstract observations (frequency space as conceptual metaphor to Cartesian space). This repetitive process is used extensively in the making of sophisticated scientific concepts. As a result, human conceptual systems assume an organic, unfolding character, where much like a growing plant each new set of leaves (new set of concepts) supports the unfolding of new leaves and structures. Notice that throughout the development of these concepts, from the model of space developed to represent human experience, to the abstracted notion of frequency space, imagination is required to construct meaningful abstractions and conceptual metaphors. This particular way of formulating novel, sophisticated, useful conceptual systems has been referred to as an *imaginative rationality*.[6]

In contrast to the inherently reductionist, mechanist, materialist image often ascribed to the physical sciences, we can deduce from the above examples (which are mainly taken from physics) that imaginative rationality appears to be quite widely used and accepted in the development of scientific concepts in the science of non-living things. In recent years the perspective of the life sciences appears to have become ossified in the thinking styles of the three -isms, especially with the embrace of a gene-o-centric perspective of living things. This was however not always the case.

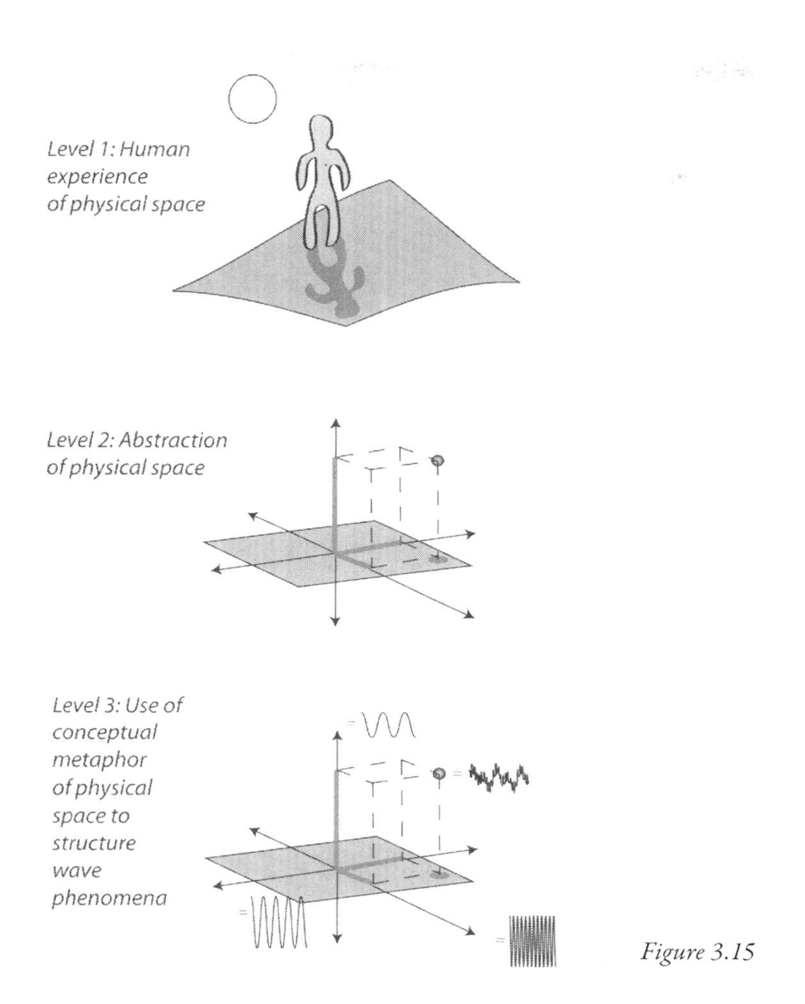


*Figure 3.15*

## *Some existing conceptual metaphors of the life sciences*

As we've discussed in preceding chapters, in the past half-century (1950s–today) the life sciences have embraced reductionism, mechanism and materialism as approaches to understanding living things. This is likely to have occurred due to the successes of these three 'ism approaches, discredit of the vitalist movement with the synthesis of organic compounds from non-living starting materials, the use of holistic ideas by non-scientific New Age communities, and even their espousal as Nazi ideology. As a consequence, imaginative rationality is almost entirely absent from the modern day life sciences. The monopoly of the three -isms is reinforced by a mandate that theories pertaining to living things be explicitly stated in terms of molecular mechanisms. This has, however, not always been the case. Looking to the past and the fringes of science today, one can find earlier, more 'primitive' understandings of living systems, which may be seeds of functional understanding warranting a degree of tolerance and further consideration. These 'outdated' concepts may be seen as conceptual metaphors resting on existing notions of the physical sciences. These may be developed into novel, functional conceptual systems to handle and provide a workable context for some of the outstanding properties of living systems.

### *Life-as-energy*

The first of these scientifically 'outdated' concepts is that of *life-as-energy*. Many have, and possibly still do, believe in an intrinsic, regulatory, ordering life-energy (or force, spark, vital essence, spirit) present in a living thing. As we have seen earlier, the concept of life-as-energy is a common belief of children, and is present trans-culturally in ancient/older systems of thought as the *qi* of Traditional Chinese Medicine, the *prana* of Ayurvedic systems, and the *élan vital* of the western vitalist movement. Of course, the concept of life-as-energy is no longer upheld by the sophisticated scientific communities of today. Rather, a living thing is seen as an immensely complex machine that produces various characteristic properties and recognized energies (heat, movement) as by-products of a metabolism driven by chemical reactions.

Consider however, that *energy* itself has no reducible components and is associated with, but not based on, material substance. Simply put, energy is a non-material, causative agent that makes things happen or stops things from happening (note that energy is discussed in more detail in Chapter 4). Material objects may possess energy as a *property* characterized by its type (for example there are kinetic, magnetic, heat, and gravitational forms of energy, among others), the quantity of energy, and the quality of the energy. Also, notice how energy is used to explain various phenomena. It is for instance understood that an object is warm because the molecules that make up the object are moving, and that the molecules are moving because they possess *energy*. It is understood that a river's flow is impeded because the dam that impedes its path is composed of bound molecules that will not break until the *energy* of the river's water flow exceeds the *energy* of those bonds. If it is very cold, a propane stove may not light because the molecules which make up the liquid fuel don't have enough *energy* to transcend the surface tension of the liquid phase to react with oxygen molecules to combust and generate a flame. If you can't get out of bed in the morning, you may say that you don't have enough energy to transcend the gravitational field of the Earth. Energy is given as the ultimate reason, the final causative factor that explains the situation, but the explanation ends there. There is no further knowing of what energy is; no further prescription of features for a type of energy to be part of reality.

Intuitively, there are many aspects of living things that correspond with the characteristics of a material form in possession of a type of energy. In the least, the idea of life-as-energy is an ideal conceptual metaphor connecting the emergent property of life to the definition of energy as that fundamental property that makes things happen. While alive, it is apparent that the life-property of the organism makes a lot of things happen. A living thing self-assembles, maintains and heals an intricate physical pattern; maintains a homeostasis that is not in equilibrium with its environment; and may think, feel, respond, act and otherwise exhibit awareness of its environment. The loss of life is consistent with the loss of 'energy', as these various things stop happening. The physical pattern disintegrates, the system comes to equilibrium with the environment, and all functions associated with the living state (thinking, feeling, responding, awareness, and so on) cease. Thus, it is apparent that the notion of energy is a befitting conceptual metaphor for various aspects pertaining to the *livingness* of an organism. Just as we may know of heat-as-fluid, we often also know of life-as-energy, and in each case the conceptual metaphor drawing the two concepts together is very befitting of observations and experience. Thus, it is not surprising that the notion of life-as-energy is found in primordial theories spanning diverse cultures, is the initial grasp of children, and was supported for a few hundred years by early Western scientists. Life and energy are intuitively and obviously connected.

Recall that the formalism surrounding the concept of heat-as-fluid did not disintegrate with the understanding that heat was also the degree of incoherent molecular/atomic motion. Similarly, even though it is well recognized that many of the properties of a living organism can be seen to arise from various underlying processes involving biochemical reactions and physical mechanisms, the idea of life-as-energy can still be tolerated as a conceptual metaphor. The idea may even prove to be useful if developed, just as the theory of heat-as-fluid proves useful even in this day and age.

### *Gaia, the living planet Earth*

Another prominent conceptual metaphor in the life sciences is that of our planet Earth as a grand living thing, where all of its landmasses, oceans, atmospheres, forests, deserts and the beings that live within, are analogous to the organ, tissue and cell systems of an organism. While the idea of Earth-as-organism was only recently (1972) introduced and developed by the scientist James Lovelock, it is actually an ancient concept.[7] Many first peoples intuitively saw themselves as but one component in a great, living, organic pattern. The ancient Greeks saw the Earth as the goddess Gaia, which is where the concept of Earth-as-organism derived its modern day name. Remarkably, although it is an age-old, simple and intuitive concept, the modern science that had been held very tightly in the grasp of the three -isms perspective was not really ready to recognize the Earth as a living thing until Lovelock's re-introduction. Unlike the metaphor of life-as-energy, the Gaia theory has not been criticized or rejected, but strangely, remains largely ignored by the mainstream scientific community .[8]

The main 'organs' of Gaia are her ecosystems filled with plants and animals (called the biosphere); the gases that the plants and animals breathe in and out (the atmosphere); the water contained within the oceans, the clouds, the soil, and the plants and animals (the hydrosphere); and the rocks, minerals and molten lava of the Earth's crust and liquid interior (the geosphere). Just as a living thing maintains its various states and rhythms (a periodic heartbeat, consistent body temperature, the concentration of salts in the blood), so too does Gaia maintain her states and rhythms. Life in the oceans requires a fairly consistent salt concentration of approximately 3.4 % (no life is possible for salt concentrations exceeding 6%). Gaia's coral reefs regulate the concentration of salt in her oceanic 'blood' at the desirable 3.4%. Also, a careful balance of 20% oxygen and 5% carbon dioxide gases is also maintained by plant and animal systems on Earth, which is essential to their own life supply. If oxygen levels were to exceed 25% fires would plague the Earth due to oxygen's flammability. Conversely, when carbon dioxide levels rise in the upper atmosphere they act as a layer of insulation which warms the Earth and makes climate patterns chaotic, establishing unfavourable conditions for life (the greenhouse effect and climate change). Gaia's temperature is also maintained at this life-favouring, cooler level than it would be as a lifeless planet receiving the sun's radiation, by the once expansive stands of tropical rainforest of South America and the other forests of the world, which suck-up the insulating atmospheric carbon dioxide. To see the Earth-as-organism is useful as it allows for a comprehensive and elegant account of the various interconnected, dynamic systems of the Earth and their delicate balance.

To assume the perspective of the Gaia concept is to assume an entirely new relationship with the Earth, and in doing so, to create an entirely new basis for 'rationality'. If the Earth is an organism, then where does that place human beings in that organism, with our tendency to deforest vast regions, stripping them to the bare soil; to create expansive deserts of concrete and asphalt, killing the life-giving soil; to cut into the muscles of her back in open-pit mining excavations and mountain-top removal operations; to poison her land, air, and water with toxic fumes and wastes; to empty her oceans of fish, while crushing extensive regions of regulating coral reef in the process; and to release excess carbon dioxide to the atmosphere in our burning of wood, coal, petroleum, and natural gas? Almost all of our agricultural, industrial and urban activities interfere with Gaia's regulatory systems in some way. Humans are breaking the rules of life. Along the lines of this metaphor of Earth-as-organism, human beings in our modern civilizations are akin to invasive parasites, consuming the resources of its host; or an autoimmune disease like lupus, in which some cells of the body fail to recognize what they are a part of and begin attacking and destroying the tissue and organs leading to death of its whole self; or like a malignant tumour, growing out of control and consuming the resources of its host, weakening and killing it. From the perspective of Gaia, the actions of humanity are hardly rational and ultimately self-destructive. Perhaps the most significant aspect of the Gaia concept is the relationship that it helps us to cultivate with the greater whole in which we are a part.

## *Biological fields*

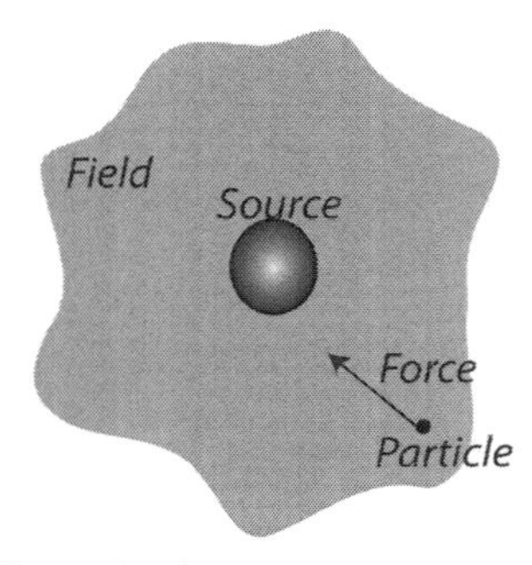


*Figure 3.16*

Another concept that faded away to near obscurity in the wake of the three -isms' revolution was that of a field associated with biological development. In the physical sciences, *fields* describe spatially distributed zones of influence executing precise and quantifiable effects on entities within their zone of influence (the basic idea is shown in Figure 3.16). A field typically arises from a source, varies in predictable ways in the space around source, and has specific effects (normally, it exerts a force) on entities that are susceptible to the field. Most physical fields are entirely imperceptible (colourless, formless, and so on), and are determined solely by observing their predictable effects on matter in their zone of influence. Let's examine in a bit more detail the nature of these fields as irreducible, non-material based entities which are accepted as a foundation concept in physics. We will then illustrate their use as a model to structure a conceptual metaphor of the developmental processes (and other attributes) of living things.

An energy field typically arises and emanates from and about a source. For gravity, the source is matter itself. The energy field spans an area of influence in space that is related to the size and geometry of the

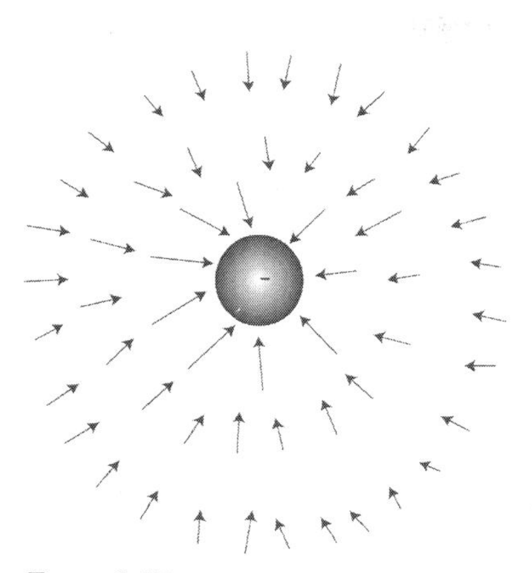

*Figure 3.17*

source, and spans space in different ways depending on which kind of energy field it is. The strength of the energy field generally falls off at distances away from the source. All possible forces which might act on a susceptible entity at various positions in the energy field can be used to represent the energy field as a *force field* (Figure 3.17). A force field is represented as a space of arrows, where the strength of the force felt by an entity at a particular position in the field is represented by the size of the arrow, and the direction of activity is represented by the direction of the arrow.

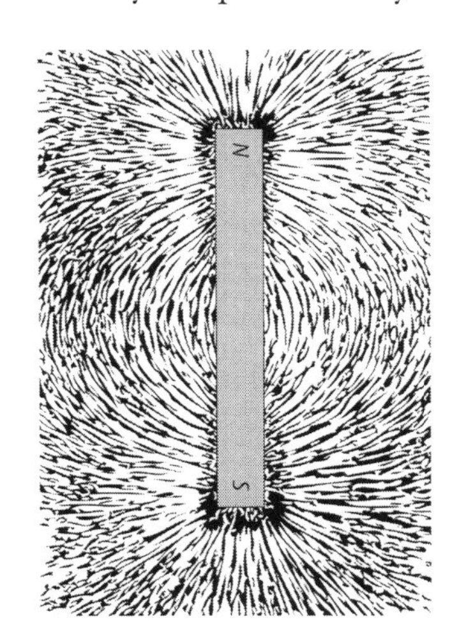


*Figure 3.18*

Common fields are the electric field about a stationary charged object, and a magnetic field about a stationary magnetic object (Figure 3.18). The force field about a magnetic object has more complex spatial properties than the gravitational field about the Earth or the field about a small charge. Every magnet has positive and negative (or north and south) poles corresponding to its endpoints. The field about a magnet causes magnetic objects (such as iron filings) to align in its presence. The alignment of iron filings allows for the magnetic force field to be visualized.

Faced with explaining the intricate self-assembly of biological forms, the early embryologists (scientists who study embryos) of 1900–1940 borrowed the concept of a physical field to account for the development of biological form.[9] These early embryologists observed a number of phenomena which were consistent with the properties and features of physical fields. Primarily, they noticed small clusters of cells (today known as organizing regions), which if transplanted to other regions of developing tissue, would produce the same tissue system (such as the limb of a salamander or the eyes of a fly) in the transplanted region, as they would if left in place. These organizing regions could be split to produce identical copies of the tissue system. On account of these observations, the organizing regions were thought of as the source of the 'field'. The spontaneous, consistent self-generation of the biological forms from simple spaces appeared congruent with the notion of a physical field influencing the properties of matter in the space around it, analogous to the orientation of iron filings around a magnet. In addition, there appeared to be a distinction between different poles of the developing system, and thus, to the organizing 'field'. This is an observation that parallels the features of a magnetic system. For example, in experiments with planaria (a type of flatworm) if the head of the organism was removed, the remaining cells would regenerate the head; if the tail was removed, the remaining cells would regenerate the tail; and if both the head and tail were removed, the cells at the former head and tail positions would regenerate both head and tail. A similar thing happens in the bar magnetic when it is cut. When cut from the south-pole side, a new south pole will regenerate at the top of the remaining piece. If cut from the north-pole side, a new north pole will form in the remaining piece (this is all shown in Figure 3.19).

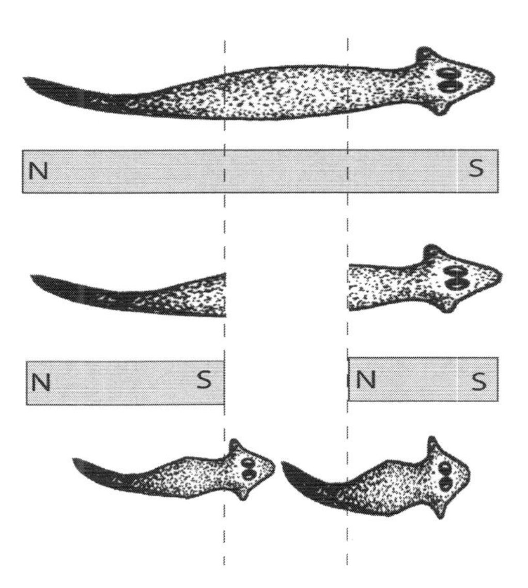


*Figure 3.19*

These above observations are all consistent with the idea of an underlying field that denotes information over a long-range to cells in its zone of influence. Notice that both traditional physical fields (magnetic, electric), and the proposed fields underlying biological development, can be seen to arise from a source (for the case of biological development, this is the transplantable organizing region), to extend their influence through a region of space (the developing tissue), to have predictable effects on form (manifestation of specific features in specific regions), and to exhibit polarity (both biological systems such as the planaria and the north and south poles of the magnet are polar systems). These biological fields are now most commonly referred to as *morphogenetic fields*, which is the term we'll use in this book. From the perspective we've been developing in this chapter, morphogenetic fields can be interpreted as a conceptual metaphor connecting the idea of a physical field to observations of biological development.

## *Morphogenetic fields and modern mindsets*

*He mentioned his field theory only tentatively 'being afraid,' he said, 'to infect you with these heretical ideas.'*
**~ Lev Beloussov relating Alexander Gurwitsch's comments from 1954**

Morphogenetic fields have largely gone the way of life-as-energy, shucked out of the life sciences in favour of viewpoints embodying a more materialistic literalism. In fact, looking to the knowledge-base of modern developmental biology, some explanations for the observations of the early embryologists in terms of the material based parts of an organism (cells and molecules) can be identified. In my opinion, rather than discrediting the notion of morphogenetic fields, this allows for the establishment of connections between the field concept and the three -isms styled perspective. As is the case for the phenomenon of heat, there are (at least) two completely different perspectives that can be assumed for biological development, and each perspective may come in handy depending on the situation where it's to be utilized. Let's now take a moment to consider some of these connections to the three -isms perspective.

As we've seen in Chapter 2, a developing organism can be seen as a growing collective of cells that are assuming characteristic forms and functions as they move on to differentiation states. Each cell of the developing organism contains the same genetic information, and specific sequences of this genetic information (genes) code for the production of specific proteins. Self-produced proteins (morphogens) act on exposed cells to change their differentiation state by changing genetic expression. Through various mechanisms, a pattern of morphogen might be established throughout the cellular collective, thus generating a corresponding pattern of differentiated cells. Under this cell, gene and morphogen perspective, the organizing regions may form as localized, specialized areas of cells via these various pattern-making strategies. Once created, it is believed that these organizing regions exert their 'field like' effects by emitting further morphogenic substances in additional pattern making schemes. Thus, although the details of these mechanisms still haven't been sorted out, it appears that the observations once attributed to a morphogenetic field can also be described in terms of a growing set of interacting cells, their genes, and self-produced signalling molecules.

While most concretely this notion of a morphogenetic field may be interpreted as a literal chemical pattern, it may also be attributed to the inherent nature of complex systems, where algorithms such as the chaos game provide an intuitive illustration of how intricate, reproducible structures can arise as emergent patterns from an underlying set of relatively simple rules, even when played out with a random selection of events. As we've seen, the chaos game demonstrates that for even a simple set of instructions with a random progression of events, an intricate structure with long-range features can be reproducibly generated. The chaos game outlines the importance of the relationship between the existing system and new elements forming in the context supplied by the existing system. Also, it's clear that the final structure is an attractor pattern to which the system converges. As we saw earlier, the chaos game reproducibly manifests the Sierpinski structure, which appears in the arrangement of points almost as if new points are drawn in towards (attracted to) the geometric structure, seemingly analogous to the orientation of scattered iron filings in the field of a bar magnet. Cognitive parallels can also be drawn to the process of biological development, where intricate, long range structures reproducibly manifest out of a homogeneous cluster of growing cells, seemingly influenced by some sort of overriding 'field'. Therefore, an additional interpretation of the morphogenetic field is as the attractor pattern to which the system converges due to the activity of a dynamic algorithm involving cells, molecules and a developing context, which is the newly forming organism itself.

However, the Ukrainian embryologist Alexander Gurwitsch, who was among the pioneers of the biological field concept, believed quite literally in a real-world physical field in biological development.[10] Gurwitsch believed that biological fields were not associated with a chemical messenger, but rather, were of a similar nature to the fields that physicists of the day were working with (electromagnetic fields). Thirty-five years examining embryonic development only refined and reinforced Gurwitsch's belief in an underlying physical field assisting with the organization of living forms. Early in his career Gurwitsch writes: 'The field acts on molecules. It creates and supports in living systems a specific molecular orderliness. This means, in our opinion, any spatial arrangement of the molecules which cannot be derived from their chemical structures.' By 1944 Gurwitsch believed that: 'A cell creates a field around it, that is to say, the field extends outside the cell and into extracellular space. Therefore at any point of a group of cells there exists a single field being constituted of all the individual cell fields. Hence, the properties of this aggregate field will depend, besides other factors, also on the configuration of the multicellular whole.'

So, we have three possible interpretations for a morphogenetic field. The first is as the literal pattern of a morphogenetic chemical, the next as the attractor pattern of a complex, non-linear process, and finally, we could take the fields literally and look for the existence of a physical field in developmental biology. In this book we don't concern ourselves too much with the underlying mechanisms, but instead highlight the field as a conceptual metaphor and set out to see what we can do with the idea.

## *Rupert Sheldrake's morphic fields*

While the notion of fields in biological development have fallen out of fashion in the mainstream life sciences, the former Cambridge biologist Rupert Sheldrake has more recently re-invented and extended the idea.[11] Sheldrake argues that reductionalist models of biological development (genes and protein synthesis) are not enough to account for the complex, precise, and coordinated processes of living things. Instead, Sheldrake believes that entities he calls *morphic fields* exist and act to directly specify the development of biological forms as a function of space and time. Morphic fields are a term that encompasses not only morphogenetic fields, which pertain to biological development, but also other kinds of field relating to a living thing's perception, behaviour, and knowledge-base. Thus, Sheldrake's morphic fields are involved with informing and directing inherited instincts and processes of consciousness itself.

In Sheldrake's model, morphic fields have no physical basis, and genetic material is viewed as a passive receptor on which the morphic fields act. It is the morphic field that turns genes on and off in the right cell, in the right place, and in the right time. In biological development this means that the features of the morphic field act on cells and their genetic material at certain regions of space and points in time, orchestrating the formation of specific features in precise locations. This is how Sheldrake explains the relative lack of diversity in genes and gene number, in contrast to the immense diversity in life forms. It is the morphic fields that are the inherited, causative agents responsible for the biological forms and behaviours, and not the genetic material itself. The genetic material is simply a receiver for the causal action of the morphic fields. Therefore, two entities with the same genetic material could develop into entirely different organisms if they were each associating with different morphic fields.

In Sheldrake's theory, consciousness is attributed to a morphic field associated with the brain, whereby, similar to genetic material, the brain acts as a physical transmitter and receiver of the morphic field. This morphic field of consciousness is thought to extend far beyond the physical body. Sheldrake suggests that when we see, our mind's morphic field actually extends all around us, reaching out to 'touch' the space of the world we are visualizing. In addition, Sheldrake suggests that a living thing may tap into a collective morphic field that belongs to the species itself, as a non-localized bank of collective knowledge. Sheldrake posits that it is the existence of collective knowledge, and tapping into it that accounts for some unexplained phenomena of nature, such as the individual-generation, lengthy, precise migration of monarch butterflies, as well as telepathic and extrasensory experiences in humans and other animals.

Sheldrake's morphic fields are free-floating entities without a material basis, a formal representation, or a mechanistic explanation. Lacking this material and mechanistic basis, morphic fields are tainted with the problem of being scientifically unfounded as they are inherently not accessible to experimental tests that focus on matter-based cells and molecules. Thus, Sheldrake's ideas have been heavily criticized by mainstream academia which demands explanations in terms of the three -isms perspective. Sheldrake currently studies the effects of these proposed morphic fields, primarily in mediating telepathic and extrasensory experiences between animals and humans, but on account of the base rejection of the notion of morphic fields, his ideas remain largely ignored by the formal academic community.

But what if we think of Sheldrake's morphic fields as a conceptual metaphor continuing and expanding the biological field concept of the early embryologists to account for even more observations for things of a living nature? How does this fresh perspective make the idea any more scientifically palatable? Well, although Sheldrake's morphic fields, and the preceding concepts of morphogenetic fields and life-as-energy, self-consistently match up with a number of real-world observations of living things, they are debunked by default for having no explanation in terms of a mechanism involving reducible material parts. Sheldrake and associated morphogenetic field supporters could go on demonstrating the functional parallels between field properties and living forms until they're blue in the face, but without a three -isms explanation or connection, they will in all likelihood be ignored. However, in this chapter, we've pointed out that many axiomatic concepts, mostly from the physical sciences, are based only on the functional parallels between the less-familiar and the more known situation. Oftentimes, there are both reductionist and holistic (conceptual metaphor based) interpretations of the same phenomenon, where each interpretation is useful depending on the situation. If this is acceptable in physics, then we can see the obvious place in science that ideas such as biological fields and life-as-energy fall into. They are holistic concepts based on befitting conceptual metaphors. As conceptual metaphors, these two notions are no less artificial or acceptable than their widely accepted foundation concepts of energy, fields or heat-as-fluid in the physical sciences.

Merely tolerating the concepts of a biological field and life-as-energy as befitting conceptual metaphors for various aspects of living systems is a first step in re-invigorating the life sciences with imaginative rationality. The next is to determine if concepts can be expanded to develop viable, useful new perspectives and avenues for the scientific study of living systems. As an example of how imaginative rationality might work in the life sciences, a thorough expansion of the morphogenetic field concept is presented in Chapter 5. Yet, as this development and expansion of ideas will require some slightly more detailed discussion of ideas from physics, I feel it essential, dear reader, to remind or inform you of these background concepts before we begin. Therefore, perhaps you would care to accompany me on a casual stroll through the mind garden of physics?

# 4. A Walk in the Mind Garden

*In my garden there is a large place for sentiment. My garden of flowers is also my garden of thoughts and dreams. The thoughts grow as freely as the flowers, and the dreams are as beautiful.*

~ **Abram L. Urban**

Let's imagine that scientific knowledge — what we've come to call our objective facts — constitutes a magnificent mind garden of sophisticated models that structure, with great success, many of our experiences as embodied minds in a dynamic world of interacting forms. In Chapters 5 and 6 we'll see that by actively enabling imaginative rationality through conceptual metaphor, these various conceptual delights may be plucked, chopped, boiled, brazened and baked together, synergistically creating delicious new dishes of understanding.

But first, let's embark on a gentle stroll through the garden, sticking our noses into the blossoming petal bundles, running our fingers across the waxy, smooth, rough, prickly parts, and taking in the many colours, sights, sounds, and manifestations. We will begin with some of the most basic ideas behind much of science by getting you to think about very basic models for stuff such as particles, waves, and fields. We'll then get rolling to look at systems of thought designed for various purposes, such as quantum mechanics, designed for the realm of the extremely small; and thermodynamics, designed for predicting energy allocation and spontaneous processes in many different kinds of systems based on collections of many elements. We'll end up with a discussion of energy and its many manifestations. All the while, we'll be focussing on the underlying images and conceptual metaphors used to structure these ideas and thought-systems. This journey will supply (or remind us of) the necessary background concepts which we will draw upon in further explorations of life and things living.[1]

## *Three modalities of matter and energy*

The particle, the wave, and the field are three distinct mental models commonly used to describe matter, its associated energy, and the dynamic interplay between matter and energy.

### *The particle*

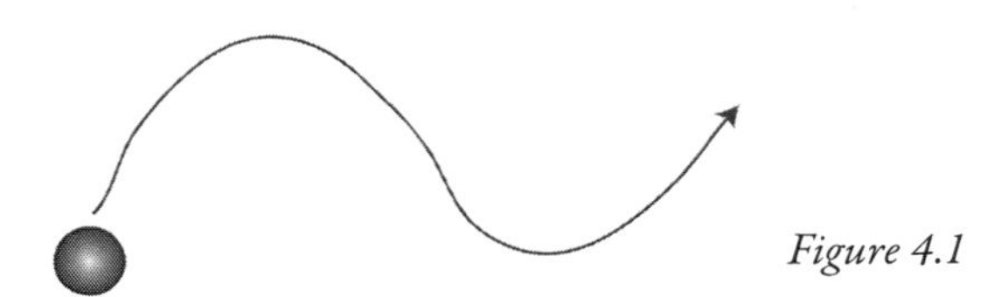

*Figure 4.1*

Sometimes we need to think about energy as an invisible thing associated with, and thought to be carried by, discrete localized forms such as rocks moving through space. We refer to these discrete, localized forms as *particles (*Figure 4.1). When particles carry kinetic energy, it's exchanged directly in collisions between particles (like two marbles hitting one another), or dissipated through interactions such as rubbing or compressing, which generate incoherent energy (heat). Particles follow distinct paths through space and time.

### *The wave*

Under other circumstances we may need to think about both matter and energy as things spread out through space. Waveforms are manifestations of non-localized patterns that carry coherent energy with them (Figures 4.2 and 4.3). Unlike a particle, a waveform's influence spreads out to fill an entire space. A waveform's pattern consists of a repeated common unit that is translated in space and/or time, which makes the

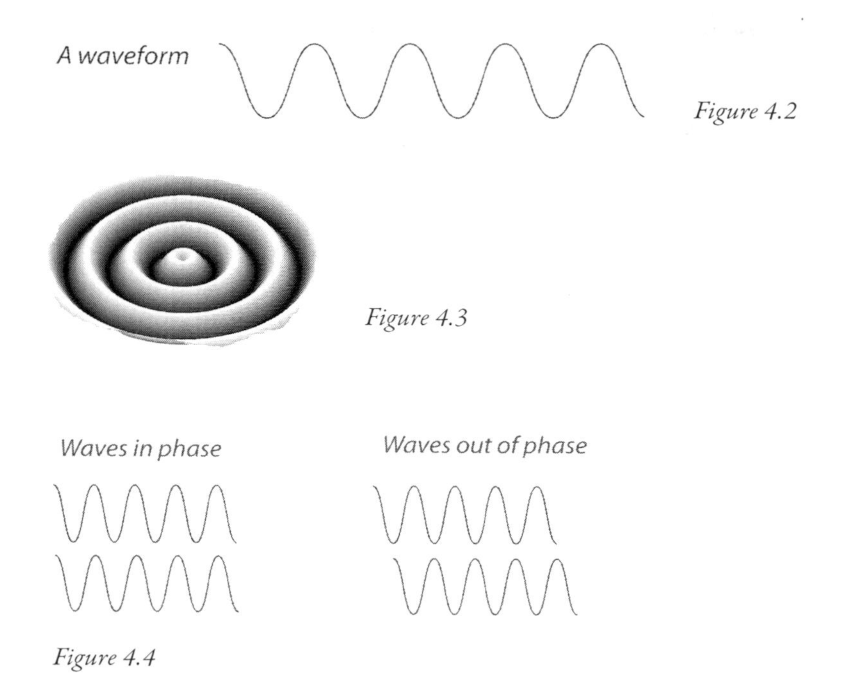


*Figure 4.2*

*Figure 4.3*

*Figure 4.4*

waveform *periodic*. Features that we call the *amplitude*, the *wavelength*, the *frequency* and the *phase* are used to describe waves. The amplitude of the wave relates how big the oscillations are in the repeating pattern unit. The wavelength defines the length of the repeated pattern unit of the wave. The frequency tells how many repeat units are present in an interval of space or time. The phase describes where the wave is in relation to another similar wave (Figure 4.4). Waves that are in phase with one another will be at the same point of their repeating cycles all along, whereas waves that are out of phase with one another will be at opposite points of their cycles.

The interactions between waves are highly influenced by the phase of the waves. Waves of similar frequencies that are in phase with one another add up to make a larger wave and are said to *interfere* constructively. Waves that are out of phase with one another cancel each other out and are said to interfere de-constructively. These constructive and deconstructive interference phenomena can result in characteristic diffraction or *interference patterns* when two waves interact.

If a waveform is supported within some container, other phenomena may arise (Figure 4.5 illustrates the whole scenario). The boundaries of the container are conditions which define what is possible for the wave action. As the wave is a non-local entity, a limit supplied at even one point will affect the nature of the whole wave. As an example, consider a rope secured at both ends to a wall. When the rope is plucked (transferring kinetic energy to it) a wave will be supported by the rope. As the ends of the rope are tied to the wall, these must remain still, and must therefore correspond with the perpetually still points of the wave. The consequence of the requirement for lack of motion at the secured ends of the rope is that any wave that the rope supports must have a number of cycles that will fit evenly into the space between the two walls. This ensures the secured ends of the rope will always be the points of zero wave motion.

As the number of repeated wave cycles that fits between the walls defines the wavelength and spatial frequency of the total wave, by securing the ends of the rope, there is suddenly a restriction on the spatial frequencies and wavelengths that are possible for the rope. These allowed states are called the *harmonics* and are whole number multiples of the so called *fundamental harmonic* which is half a wavelength and equal to the space between the walls (see Figure 4.5).

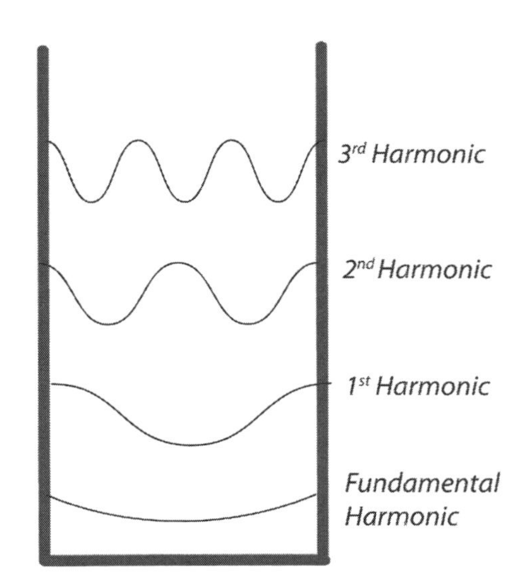


*Figure 4.5*

If a wave with a frequency corresponding to the harmonics of a container is incident upon the container, a vibration will occur within the container. This is known as *resonance*. Interference, harmonics, and resonance are fundamental characteristics of wave phenomena. It is not possible to conceive of a particle forming an interference pattern, generating harmonics in a container, or of two particles resonating with each other. This is because particles as we know them don't have a frequency, wavelength, or phase!

Some waves, such as a wave on a string, a wave on water, or a sound wave in air, are supported by a collection of particles (Figure 4.6). The energy associated with these 'stuff-based' waves is kinetic energy. The amount of energy associated with the stuff-based wave is related to the amplitude of the wave. The amplitude is a continuous thing under all circumstances, and can take on any level within limits; therefore, the energy of a stuff-based wave is always continuous.

Other wave phenomena such as light, do not need to be supported upon some medium, but are intrinsically wavelike. These are 'stuffless' waves as they do not require a supporting substance, such as rope, water or air. Stuffless waves are also called *radiation*, which refers to a wide range of phenomena including radio waves, light waves, and X-rays. The crucial difference between a stuff-based and stuffless wave, is that the amount of energy of a stuffless wave is related to its

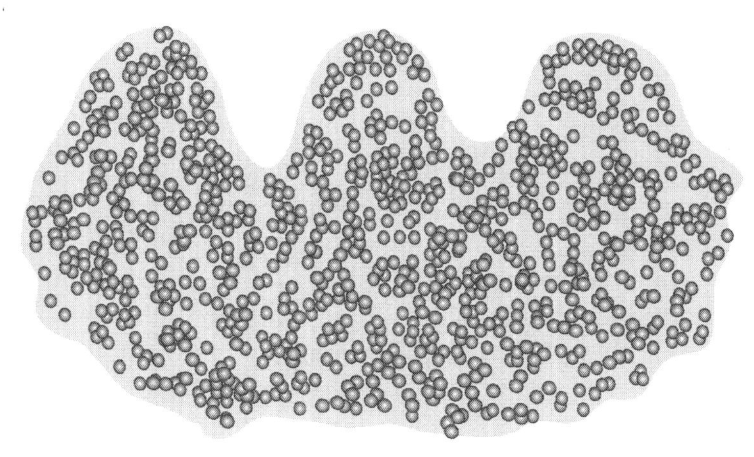

*Figure 4.6*

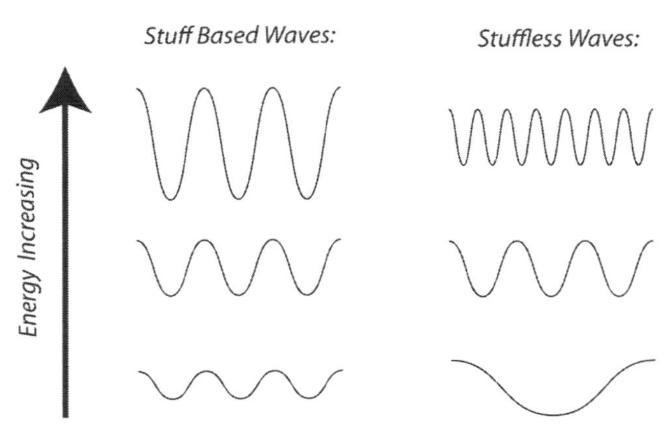


*Figure 4.7*

frequency/wavelength, instead of its amplitude (as detailed in Figure 4.7). Stuffless waves of higher frequency (or shorter wavelength), represent a wave with higher energy. This means that the pattern that the stuffless wave exhibits corresponds directly with the wave's energy. If a stuffless wave could be held in a container (they can, only the containers are very small), only certain pattern states would be supported, analogous to the case of a stuff-based wave in a container (refer back to Figure 4.5). Since the pattern state (frequency or wavelength) is directly related to the amount of energy the wave possesses, if the stuffless wave exists as harmonics, only specific energies of the stuffless wave are possible. This nature of stuffless waves in containers brings us to the important concept of the *discrete energy state.*

The concept of an energy being coupled to a wave's pattern state is in conflict with our everyday experiences, and our corresponding conceptualizations. We are used to experiencing the energy of a wave in terms of its amplitude, which we know to take on *continuous* values under all circumstances. When small water waves rock a floating dock they do so to a much lesser extent than an enormous swell, and intuitively we know (that is, it is our common experience) that this occurs because the enormous swell carries more kinetic energy in its bigger amplitude. It is the same thing for earthquakes, and loud sounds — the bigger the amplitude of the wave, the more energy it carries. Moreover, the idea of a discrete form, such as a rock, is an everyday experience; however, the idea of a discrete energy is not. Everything previously experienced in the day to day reality assumed continuous levels of energy. Our conceptual realities expanded after our observations and conceptual integration of stuffless waves.

## *The field*

A last archetypal model is the idea of a field, which we've discussed previously (see Chapter 3, pages 98–101). In physics, a distinction has been made between *active* and *potential* modalities for energy. If you hold a big rock up over the ground, we would say that it has *potential* energy because if we let go of it, it'll transfer its gravitational potential energy to an *active* energy of motion called *kinetic energy.* Another way to interpret the exchange between potential energy into active energy is to see that nature aims to minimize a build-up of any kind of energy by manifesting it in an active form. A ball will roll downhill. A sealed pot filled with water and heated forms a highly pressurized system that can explode if heating is continued. Another feature of the potential-active energy conversion is that energy can transform between these different types of energy, but the total amount of energy in any form always remains the same. Energy can never be created or destroyed.

Potential energies can be represented in more than one way (Figure 4.8). As we saw previously, potential energy is often represented by topography, especially when discussing the *stability* of a system, which means we are trying to predict if and how it may change (Figure 4.8, top). Using this model, potential energies are represented as a landscape, where unstable points are represented by peaks and stable points represented by wells. Using another image, potential energy can be understood to exist as a *field,* which is a non-localized expanse of influence that is imperceptible (formless, colourless, tasteless, unscented) except through observation of the precise and predictable effects of the field on the entities of physical reality on which it acts (Figure 4.8, bottom). The connection between energy fields and an entity under the field's influence is *force.* The force acts on an entity in the energy field, instilling the entity with active energy if it is free to move and therefore

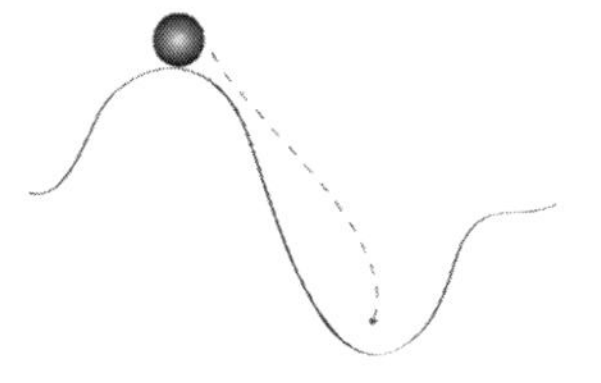

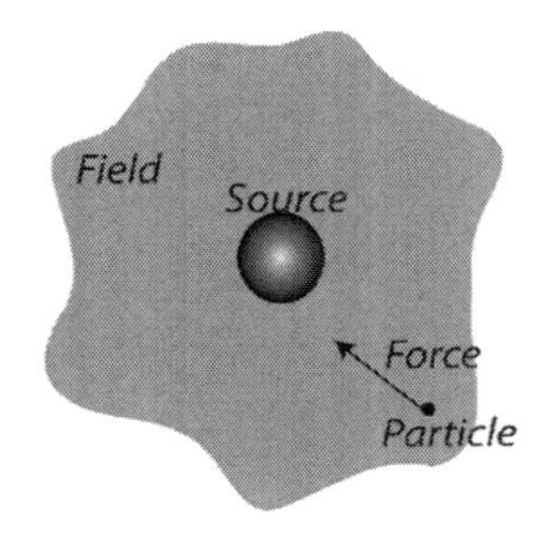


*Figure 4.8*

not balanced by other forces on the particle. The field can be represented in its potential energy form (where it's aptly called a potential energy field) or as a force field, where it is represented in terms of forces that would exist if field-susceptible entities were to be at all of those places in space where the field acts. Fields tend to arise from a localized, stuff-based source, and extend ephemerally through the surrounding space in characteristic ways. Fields are literally a depiction of energy spread through space and time.

Common fields are electric and magnetic fields. Electrons are components of atoms representing a fundamental unit of electric charge, where one electron is conveniently defined as one unit of negative charge. In an atom, the negative charge of the electron is balanced out by the positive charge of the nucleus. Therefore, a negative or positive electric charge on a piece of large-scale stuff relates to an excess or deficiency of electrons. A purely electrical field arises from electrically charged matter (positive or negative) that is not moving. The electric field about an object changes with different shapes and configurations of the charged material. For instance, the electric force field coming off of a small, spherically shaped charged material is shown in Figure 4.9-A. When two pieces of matter that are oppositely charged are placed close to one another they form what's called an electric dipole, which has a characteristic electric field much different than the spherical charges on their own (Figure 4.9-B). An even more complex arrangement of electric charges is shown in Figure 4.9-C.

When electrons move, they generate an electric current, which is what we call electricity. Magnetic fields can be formed by charged matter moving as a current, and have different properties and spatial

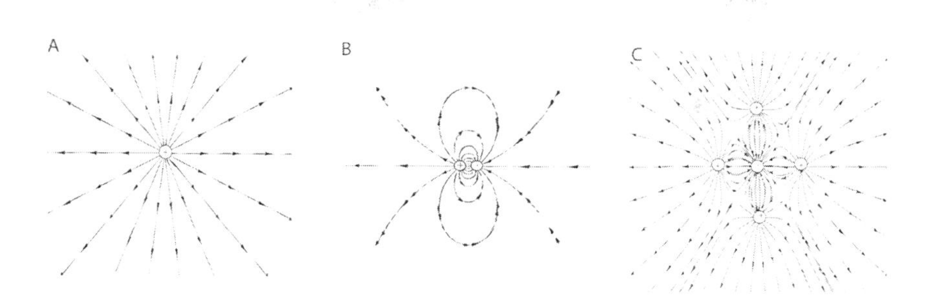


*Figure 4.9*

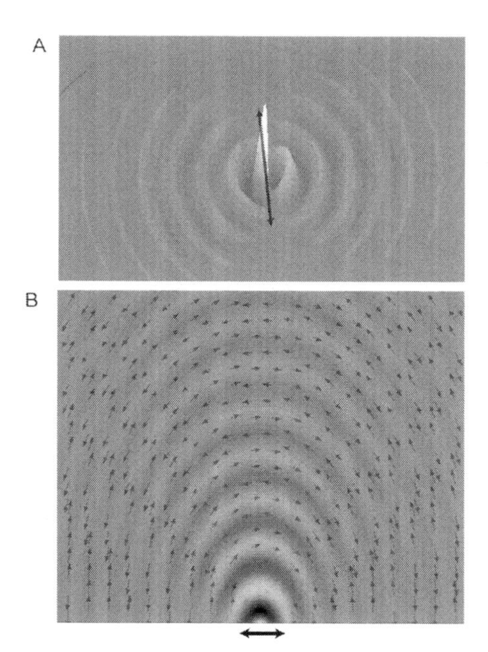


*Figure 4.10*

distributions from electric fields. When an electrical charge is *accelerated*, which means its speed changes with time, an *electromagnetic field-wave* is generated. Figure 4.10-A shows the electrical field strength of such an accelerated charge, which is a current rapidly moving back and forth in the direction of the black arrow. This field strength varies in space and time just like a wave, and even travels outwards (radiates) away from the moving charge. We call this an electromagnetic field-wave as it is also a force field changing in a wave-like way (this is shown in Figure 4.10-B). This so-called electromagnetic phenomenon is made up of both electric and magnetic fields, but for simplicity, we've ignored the magnetic fields in the images of 4.10-A and B. Light is an electromagnetic field-wave. The stuffless waves we mentioned previously are electromagnetic field-wave phenomena.

There's often a need to switch between these dramatically different models of particle, wave, and field. To work with light, for instance, one needs to interchangeably make use of all three concepts, depending on the context you're working within. Sometimes light is best described as a continuous wave with a beam-like trajectory; sometimes it's seen as localized, distinct, particle-like packets of energy; and sometimes light is seen as an electromagnetic energy field extending throughout a space and varying in time. Light is a very confusing and complicated thing to be thinking about! This is why we say that a physicist can change their conceptual view of a phenomenon as we would change a pair of glasses.

## *Life at the atomic level*

*The quantum is that embarrassing little piece of thread that always hangs from the sweater of space-time. Pull it and the whole thing unravels.*
**~ Fred Alan Wolfe (from *Star Wave*)**

Matter is composed of building block units called atoms. Scientists studying matter have demonstrated that the individual atoms are themselves composed of a central positively charged nucleus surrounded by orbiting negatively charged electrons. The atom is itself extremely small. Quantum mechanics is a system of concepts designed to describe things that are on the order of atomic smallness. Take a look at a piece of your hair and see how thick it is — there are about one million atoms that span that distance. If an atom were the size of a stadium, an electron would be a fly sitting on a bleacher! Quantum mechanics is the result of studying things this small, often under isolated situations, in which they tend to behave very differently from the stuff of our everyday experiences.

In our everyday experience, if we have some sort of solid, independently moving particle, like a rock for instance, and we set it down on a table in a quiet room where nothing much is going on, it'll probably just stay pretty much where we put it. What the rock does has very little to do with whether or not we're looking at it, how we're looking at it, or what we're thinking or feeling about it. If we close our eyes and turn our back on the rock, when we turn back around and open our eyes, chances are it'll still be there in pretty much the same place we remember seeing it last.

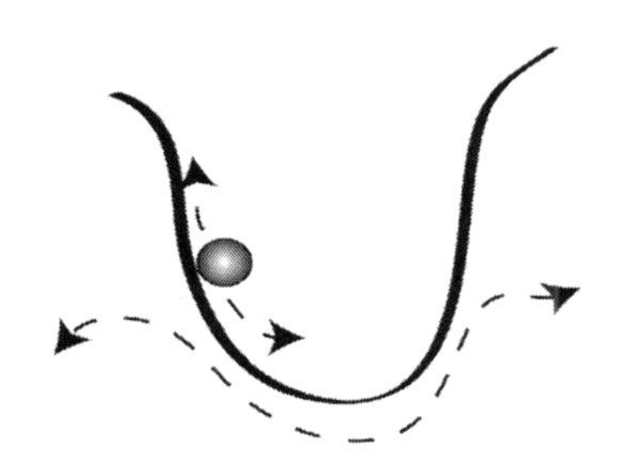

*Figure 4.11*

Let's set up a simple scenario to help demonstrate how unusual life at the atomic level is. A simple potential energy well can be made by placing the rock in a bowl (Figure 4.11). Tilting the bowl back and forth a bit would cause the rock to roll up and down the sides of the bowl. The movement of the bowl supplies the rock with kinetic energy to roll up the sides of the potential well, where it rolls back down under the influence of gravity. This rock-bowl experiment is not too exciting as it is so ordinary and consistent with day to day experiences. Unless the rock is thrown right out of the top of the bowl, it stays looking like a rock and moves around with no limits on the amount of kinetic energy (velocity) it can assume. It behaves pretty much like all of the stuff observed in the everyday human realm.

Electrons, and other things that are very, very small, are a different story altogether. Our experiences with them rattle our basic concepts such as that of the *discrete* versus the *continuous*, and the distinction between *particles* versus *waves*. In fact, the fundamental nature of electrons seems quite perplexing. Under some circumstances, they seem to behave like particles. Like other particles, electrons seem to have a distinct locus of existence, to follow trajectories, to collide, scatter and transfer kinetic energy like hard glass marbles. However, under other circumstances electrons act like waves and can make interference patterns. Needless to say, it is difficult for us to picture what the electrons are actually doing under these circumstances.

To demonstrate how strange the behaviour of a quantum 'particle' (such as an electron) is, let's consider what would happen if we had a very small electron-confining bowl, and were able to place an electron into it, instead of a rock. Like the rock, the electron seems to be a discrete and localized form. Given that electrons seem to behave very much like particles in many situations we have observed them we may expect to observe similar behaviour of the electron in the bowl as we saw with the rock. However, what is observed in experiment, and predicted by quantum mechanics theories, is a very different picture indeed. The energy of the electron in the potential well is limited to *discrete* values that correspond to wave-like pattern states of existence, analogous to the harmonics of a wave. In this situation the electron doesn't look like a particle at all, as its very existence seems to spread out into the shape of a wave pattern harmonic within the potential well.

In quantum mechanics, there is a blending of the seemingly incongruent concepts of particle and wave. This is called wave-particle duality, or the particle-wave paradox. The everyday concept of a particle

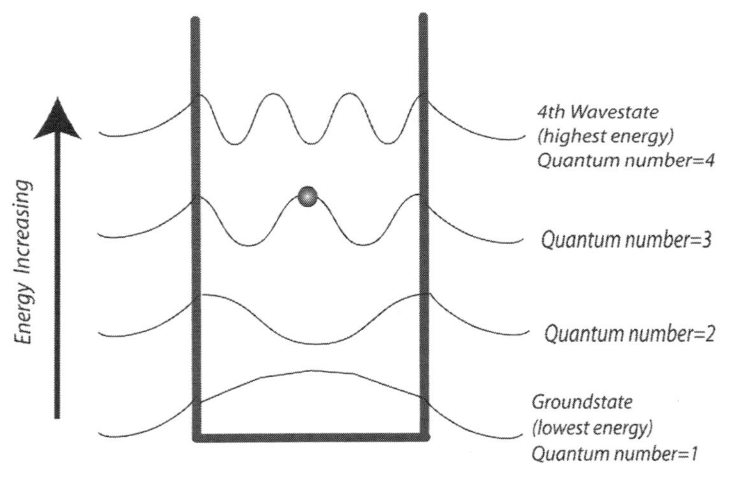


*Figure 4.12*

trajectory is replaced by that of a *wavefunction* and *wavestate* (note that this is a totally non-intuitive concept and can be seen schematically in Figure 4.12. The wavefunction contains all of the information required to describe the location and motion of a particle associated with it under all *possible* circumstances (take another look at Figure 4.12). The wavefunction produces a whole set of wavestates. The particle's wavestate refers to a particular harmonic of the wavefunction. The amplitude of the wavestate is related to the *probability* of finding the particle at a particular location and time. If the amplitude of the wavestate is large at a point in space and time, then there is a high probability of observing the particle there, while if it is zero, then there is no probability of ever finding the particle there. The wavestate represents a patterned state of existence which the electron either exists as, or is guided by, depending on the interpretation of quantum mechanics. The energy of the wavestate is the same as the total energy of the particle. If these ideas are new to you and you are confused, don't worry, it's perfectly normal. Quantum phenomena do not correspond with our everyday experiences and are therefore non-intuitive ideas for us. We have to contextualize information from the quantum realm using bits and pieces of our experiences from the everyday world as conceptual metaphors.

A basic picture for a quantum mechanic system is that of a particle in a potential energy box (particle-in-a-box) which has high potential energy at the walls and zero potential energy within the walls (see Figure 4.12). The wavefunction is affected by the boundary conditions so that in the region between the walls only states with a form which includes a complete set of wave cycles are allowed, just like in the case of the stretched rope. For this basic picture, the allowed wavestates are the same harmonics as in the case of the stretched string. Just as in the case of stuffless waves we described earlier, the energy of the quantum particle is directly related to the harmonic of the wavefunction associated with the particle. A so-called *quantum number* is used to keep track of the wavestates of the particle-in-a-box (Figure 4.12). For the case of the quantum particle in a potential energy box, the quantum number is simply the whole number multiple of the fundamental harmonic, and is directly proportional to the energy of the state.

Notice that the wavestates for the particle, which have amplitudes that are related to the probability of finding the particle at a particular place and time, extend outside of the potential well. This is equivalent to observing the egg popping right out the side of the bowl, even though there is no hole, or other structural imperfection in the bowl.

When potential energy wells take on a more involved geometry than the potential energy box, the corresponding wavestates assume more complex structures. What if for instance, instead of having a potential energy landscape shaped like a box or a well, we look to three dimensions to consider a potential energy that changes in a spherical manner? This spherical distribution could represent the case of a negatively charged electron associating with a positively charged atomic nucleus, which turns out to be an approximation of the form of the simplest atom (the element hydrogen). In this case the potential energy between the electron and the nucleus changes depending on the distance between the two entities, but it is the same relationship for any angle around the nucleus, and therefore has an isotropic quality making it approximately spherical. While it's hard to wrap one's mind around without studying it for a year or two, quantum mechanics relies heavily on a mathematical system where one puts in the geometry of a potential energy with which the quantum particle is interacting, and gets out a set of wavestates that describe the various possible states of the quantum particle and its possible discrete energies corresponding with these states.

In a spherical potential, a quantum particle such as an electron has a structural hierarchy of wavestates available to it, with each wavestate corresponding to a different spatial form. Rather than being wavelike, as they are in the example of the particle-in-a-box, these wavestates are shaped more like balloons. In spite of the fact that 'wavestate' is a non-ideal term as these states are no longer expectedly wavelike, we'll keep the original notation for consistency. The wavestates in this situation appear more complex, but ultimately still do the same thing that they do in the simpler example of the particle in a box, as they specify regions of space where the particle spends most of its time.

Russian babushka dolls, which you've hopefully seen before, open to reveal many dolls of increasingly smaller size nested within one another. This is a type of hierarchical structuring that's a good starting point to help us get our head around the nature of a quantum particle in a spherical potential energy distribution (the quantum description of a simple atom). The simple atom is composed of nested energy shells, which are in turn composed of nested sub-shells, which are in turn composed of nested wavestates, each with their own form. In this analogy, each complete babushka doll in the set represents an energy shell of the quantum atom. However, each babushka doll (energy shell) in the set has various degrees of complexity and intricacy associated with it, depending on where it is in the nesting sequence. The largest babushka doll, perhaps the queen of the babushka doll community, is the most intricate, and corresponds to the highest energy. The last one in the nesting set has the simplest form and corresponds to the least energy. Note that the energy is associated with the whole doll (whole set of wavestates), much like a family name. We can imagine that all of the little dolls within the whole doll have the same last name (share the same energy value).

For the case of the particle in a spherical potential, because of the structural hierarchy in the wavestates, three interrelated integer numbers are used to specify any particular wavestate. There is the principal quantum number, which we'll represent by the letter N, which specifies the main shell and is also directly proportional to the energy of all states that are in that shell. This principal quantum number indexes which complete babushka doll in the set we are referring to. The next two quantum numbers are used to specify how much detail is associated with that particular 'doll' (energy shell). There is the secondary quantum number, L, which specifies the sub-shells of the main shell and the tertiary quantum number, m, which specifies the actual wavestates in the sub-shells. As this stuff is actually important for the work that follows in Chapters 5 and 6, let's take a moment to see how this organization of the quantum atom might appear.

The state with lowest energy for the quantum particle in the spherical potential corresponds to N=1, with L=0 and also m=0 (Figure 4.13). This is the simplest energy shell consisting of only one sub-shell and one wavestate. This wavestate looks like a simple, spherical ball. In the

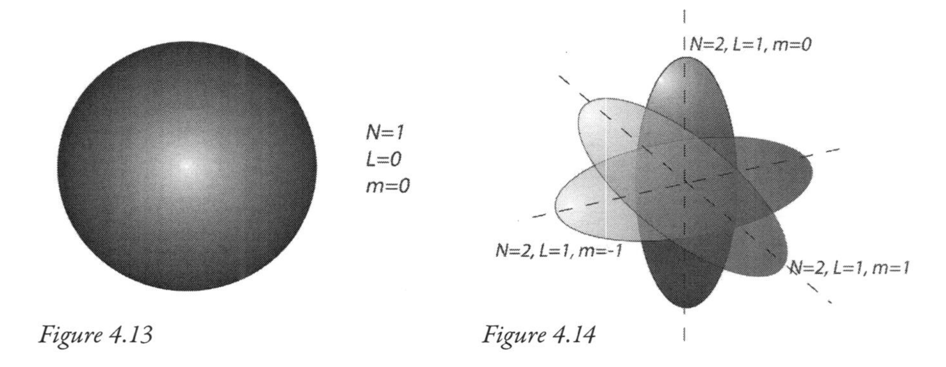


*Figure 4.13*

*Figure 4.14*

diagram image of this wavestate, the electron spends its time in the areas indicated by dark pigment.

For the next highest energy (Figure 4.14), N=2, there are two sub-shells, L=0 and L=1. Each of these two sub-shells, indexed by the value of L, contains wavestates as indicated by the value of m. For the L=0 sub-shell there is just one wavestate indexed by m=0, the form of which we have just described. For the L=1 sub-shell there are three wavestates corresponding to m=1, 0 and 1. These wavestates are each shaped like long balloons and are oriented along the three basis dimensions of space.

For the next energy level up (Figure 4.15), N=3, and L=0, 1 or 2, so there are three sub-shells each containing wavestates as indexed by the value of m. The structure of the first two sub-shells indexed by L=0 or L=1 we have already examined. For L=2 there are five values for m (–2, –1, 0, 1, 2), which correspond to five different wavestates.

These are all put together by nesting the shells within one another, as is shown in Figure 4.16.

An ideal way to handle the hierarchical structure of the wavefunction is to break it down into a table indexed by the primary quantum number N, and its associated secondary number L (as shown in Figure 4.17). Recall that while each energy shell has different levels of structural intricacy consisting of sub-shells and individual wavestates, the energy of all structures within one shell are equivalent to one another. The total number of wavestates sharing the same energy value is referred to as the *degeneracy* of the energy state. As the energy gets larger, and therefore the quantum number N takes on bigger values, there are more wavestates at the same energy and the degeneracy increases dramatically.

This rather non-intuitive and complex formalism that details the places where the electron of an atom (quantum particle in a spherical potential) hangs out has been successful as it's given us an understanding of chemical bonding between atoms, as well as insight into the reasons why the atom interacts with light, magnetic fields, and other things in the way that it does.

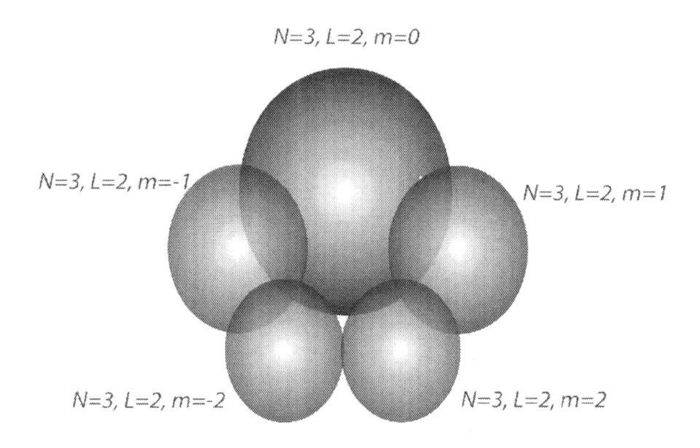


*Figure 4.15*

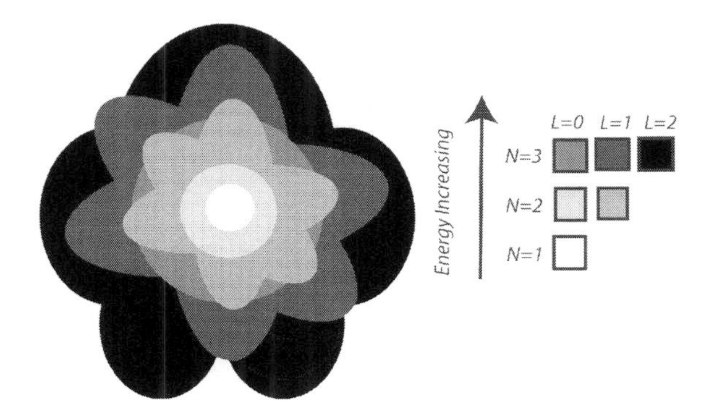


*Figure 4.16*

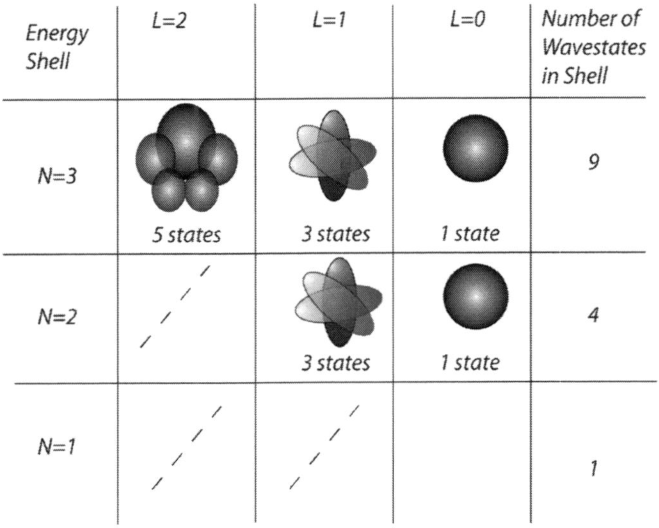

| Energy Shell | L=2 | L=1 | L=0 | Number of Wavestates in Shell |
|---|---|---|---|---|
| N=3 | 5 states | 3 states | 1 state | 9 |
| N=2 | | 3 states | 1 state | 4 |
| N=1 | | | | 1 |

*Figure 4.17*

## *The particle-wave or the wave-guided particle?*

*Why is the pilot wave picture ignored in text books? Should it not be taught, not as the only way, but as an antidote to the prevailing complacency? To show us that vagueness, subjectivity, and indeterminism are not forced on us by experimental facts, but by deliberate theoretical choices?*
**~ John Bell speaking about David Bohm's pilot wave theory**

The theory that is quantum mechanics came about after physicists encountered a number of phenomena that could not be accounted for using the ideas and rules in existence. A real-world experiment which develops our appreciation of the quantum realm is the formation of electron interference patterns.

When two waves of a similar frequency originate from two closely spaced sources, they interact as they travel through space. Where the waveforms are aligned they add up to produce a super-wave (constructive interference), and where they are completely misaligned, they cancel each other out to form no wave at all (deconstructive interfer-

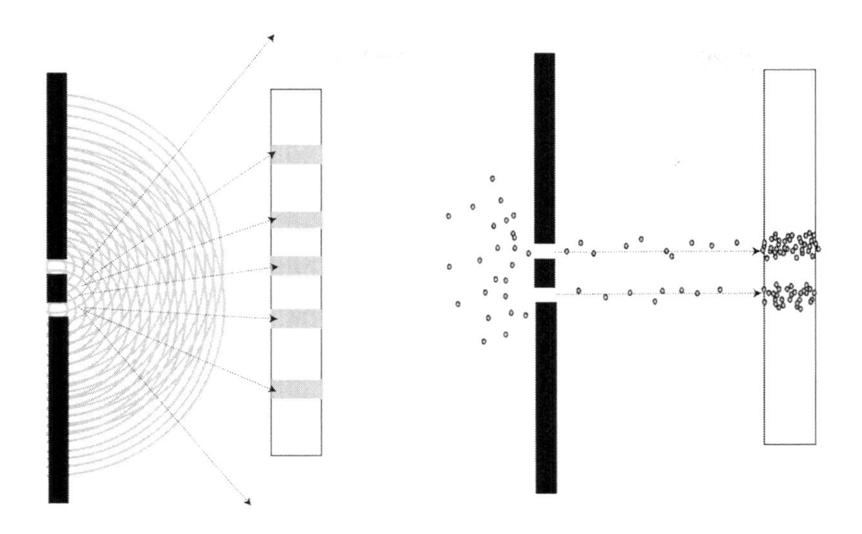

*Figure 4.18* *Figure 4.19*

ence). If a flat screen is placed at some point in their subsequent paths a so-called interference pattern, consisting of a bunch of equally spaced spots, will form (Figure 4.18). The spots represent constructive interference of the two waves, while the empty places in between represent their deconstructive interference. In our everyday worlds, interference is a property unique to waves, and is not congruent with the concept of particles. If we were to repeat the two-slit experiment by passing particles instead of waves through the two slits, we simply end up with two bands on the screen where the particles have landed after passing through one of the slits (Figure 4.19).

Now for the most part, electrons seem to be just like particles. They seem to follow distinct trajectories, to bump into one another and scatter, and to have a distinct locus of existence. Yet when electrons are passed through slits what comes out the other side is an interference pattern consisting of a large series of spots! Under the conventional interpretation of quantum mechanics (the Copenhagen interpretation) this very strange behaviour of electrons is accounted for by proposing the electron is both a particle and a wave at the same time. Under some circumstances, such as the electron diffraction experiment, the electron's existence is supposed to spread out into the shape of a wave pattern and to therefore behave like a wave during those times. This is a very hard thing to reconcile in one's mind — how something can be both a particle and wave at the same time.

A relatively well-kept secret is that while the Copenhagen interpretation is the most popular and recognized of quantum mechanics theories, there are actually a few different ways that the quantum realm has been envisaged. Working at the crucial time when quantum mechanics was being developed (the 1940s), the American-born physicist David Bohm made the 'outrageous' claim that a particle, even a quantum particle, was always a particle. In fact, if we look closely at an actual electron interference pattern (Figure 4.20) we can see that the pattern does not appear to gradually increase in homogeneous intensity. Rather, the electrons show up as distinct spots on the detection screen, just as particles would.

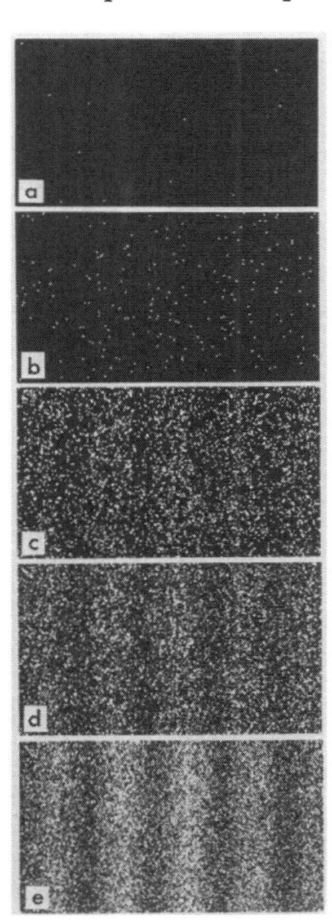

*Figure 4.20*

Bohm reformulated quantum mechanics with the concept that associated with the quantum particle is an invisible guiding waveform (the pilot wave), which influences the electron's behaviour. In his theory, which is called 'Bohmian mechanics' or the 'pilot wave theory', electron diffraction is accounted for by prescribing a guiding waveform to each electron.[2] In a collection of many electrons, their individual pilot waves combine so that the whole collective system can be associated with a single guiding wave. The form of the guiding wave is influenced over a long range of the particle's surroundings, and it is the electron's guiding wave which is diffracted through the slits. All the while, the electrons continue to be particles and follow trajectories described by the pilot wave (Figure 4.21).

Pilot wave theory accounts for the same quantum phenomena that the Copenhagen interpretation can account for, but Bohm's theory is more conceptually palatable as particles remain particles, following distinct

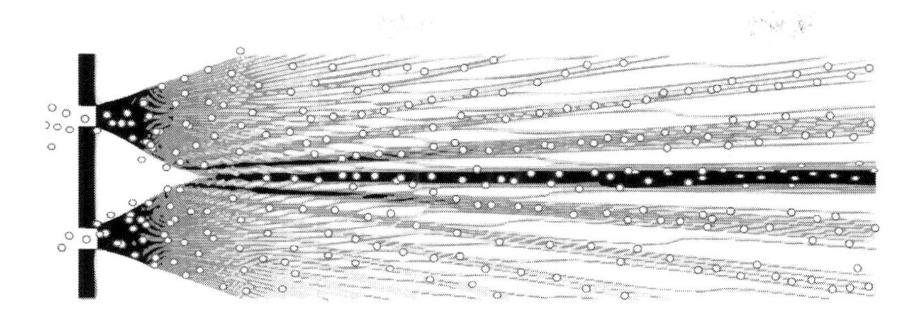

*Figure 4.21*

trajectories in space and time. Yet, for reasons that may forever remain unknown, Bohmian mechanics remains relatively unacknowledged by the physics community. It's not taught to students learning about quantum mechanics in school. It's not a hot topic of discussion or research.

I mention David Bohm's pilot wave theory here as, for our intended purposes, it's a much better model to have in mind. We're going to be using ideas from quantum mechanics to structure various features of natural systems. Natural systems have more basic parts such as cells, which are clearly behaving in fairly good accordance with the standard view of a particle. In Chapter 5, we are going to rethink morphogenetic fields and it will be helpful if they can be seen as pilot waves, acting to influence and pattern discrete particles (cells) under their influence. But before we can get to that, we need to go over just a few more concepts involving how science has handled communities of more than one element.

## *Systems of systems*

On an inquisitive walk through the mind garden, one might notice that an interesting demarcation exists in our conceptual approaches, which arises because we consider phenomena at different perspectives of scale. There is a whole conceptual approach devoted to the study of individual things on their own, and there is another conceptual approach devoted to the understanding of the collectives formed when many of these individual things come together. Each individual thing, such as an electron, molecule, photon of light, rock or a single cell, can be seen as a little system with its own properties, behaviours, influences and interactions. When large numbers of these individual elements come together as a collective, we can think of them as a 'system-of-systems' (Figure 4.22) as the collective has its own properties, behaviours and influences that are unique to it and not found in the isolated individual things on their own. Due to the large number of elements that come together, these systems-of-systems often assume macroscopic proportions, and are what we deal with in our day to day lives. An understanding of systems-of-systems is a fundamental aspect of understanding emergent phenomena, as emergent phenomena arise from the collective activity of the system-of-systems and are often related to the relationships between individual systems of the collective.

Quantum mechanics is one conceptual system designed to work with the individual systems of matter (electrons, atoms). Conceptual systems

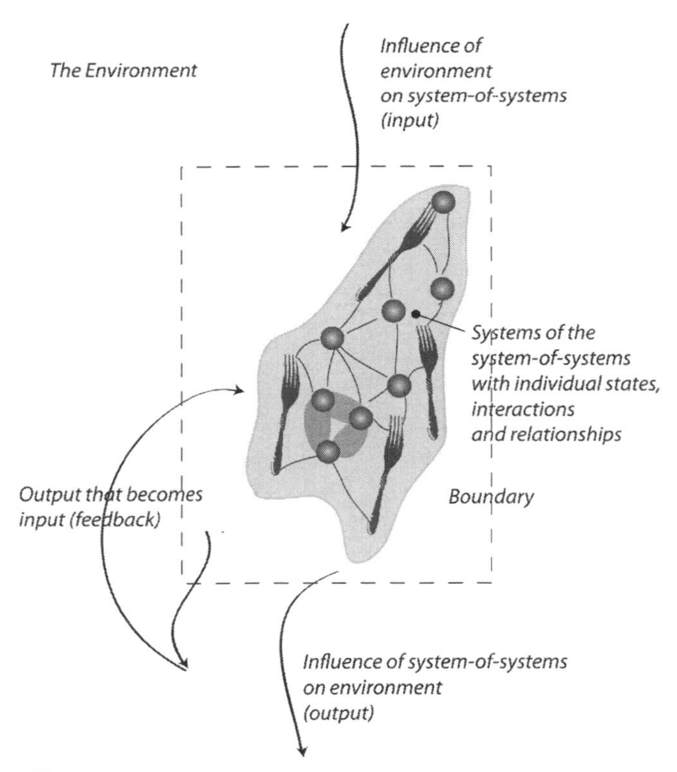


*Figure 4.22*

designed to work with the collective system-of-systems formed by these individual parts coming together are called *thermodynamics* and *statistical mechanics.*[3] Thermodynamics requires entirely new concepts to deal with the entirely unique properties and processes of the collective. These new concepts are things like entropy, temperature, and heat, and the concept that a system-of-systems has a collective state and can undergo transformations to other states. The collective state of the system-of-systems depends on the states and relationships between all of the individual components. Statistical mechanics looks to both the features of quantum systems, and the emergent properties of thermodynamics, to explain how these properties arise from the quantum viewpoint.

## *Organization, entropy and order*

We tend to conceive of organization in two ways. The first arises from our basic experiences with the physical placement of form-based objects in space. We intuitively know that we can organize objects in space. In the placement of some form in space, there may be continuous options, or there may be some additional organization or influence such as a container, which means the object can only be placed in specific, discrete locations. For instance, if we put eggs on a table (Figure 4.23), they may assume any position at all (*continuous* options for the position state of the egg) but if we put eggs into an egg-carton, they're forced to occupy set locations (*discrete* options for the position state of the egg).

*Discrete position states of egg:*

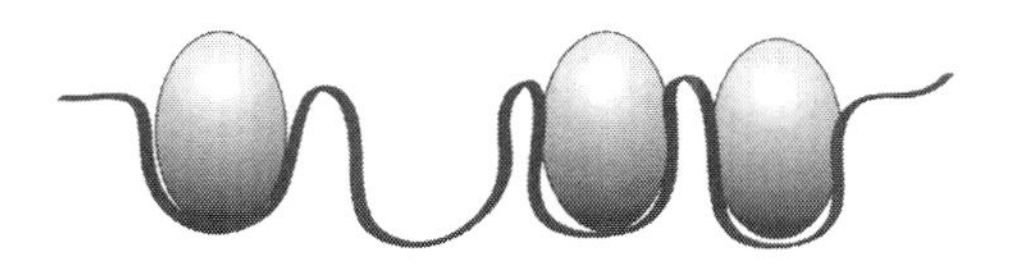

*Continuous position states of egg:*

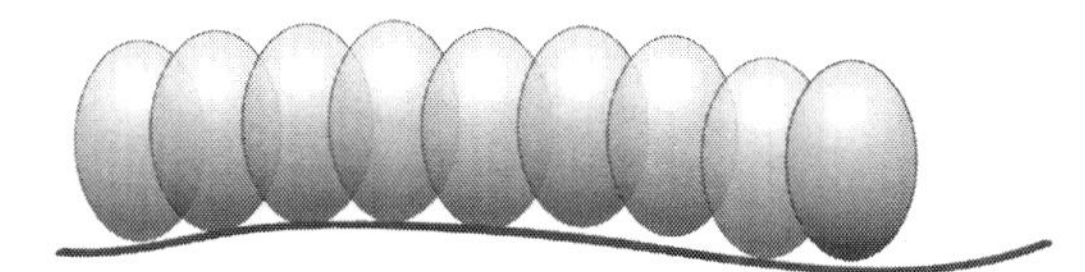

*Figure 4.23*

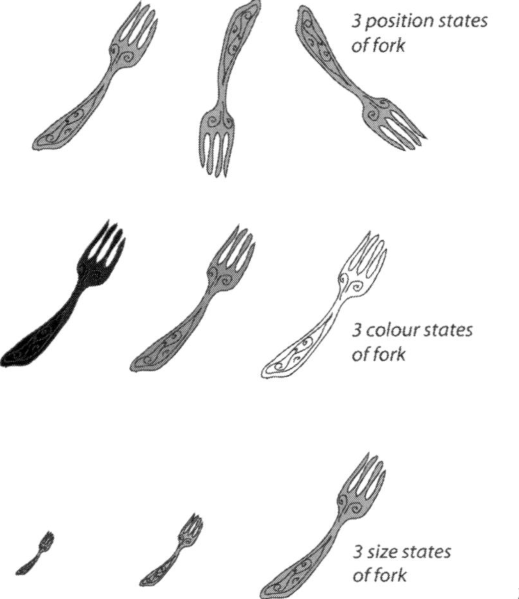


*Figure 4.24*

Using our basic experiences with the placement of objects in space as a conceptual metaphor brings us to the scientific notion of a *state* (Figure 4.24). This idea of a state reflects the amount of some property that an object may exhibit, such as colour or size. By conceptual metaphor we associate the amount of the property with a position in space. Just as in the case of actual placement of objects, the options for the state that something may exist in may be continuous or discrete.

In quantum mechanics, there are a number of wavestates which represent discrete amounts of energy that a particle may exhibit. The electron can 'occupy these states', meaning that it exists with the amount of energy and the spatial probability distribution specified by the particular wavestate. We draw the wavestates of the wavefunction as if they are rungs of a ladder, or spots in an egg carton. The particle is seen to *take the place of* the state that it occupies. It is, however, not actually taking a physical position, but assuming a particular energy and distribution. The top panel of Figure 4.25 shows four different energy states (labelled E1, E2, E3 and E4) of a hypothetical quantum system. Suppose these energy states are available to a hypothetical

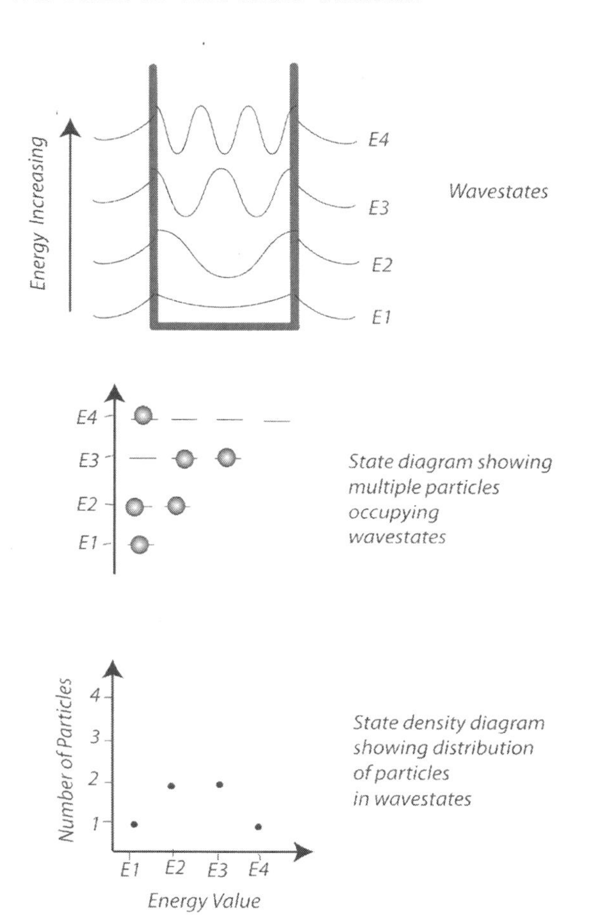


*Figure 4.25*

quantum particle, and therefore, a particle can be said to occupy one of them. In other words, these are a set of discrete energy states for these particular quantum particles.

A *state diagram* (also shown in Figure 4.25, middle) is a visual representation of the distribution of entities amongst available states. The diagram shows how a collection of six quantum particles might be distributed amongst the four possible energy states. Note that this state diagram shows that the degeneracy of the states is increasing with the energy of the state, which means there are more states (represented by open dashes) of the same energy as the energies get larger. This state diagram specifies the larger-scale state of the system-of-systems formed by the collection of quantum particles.

The final tool used to represent how a set of states are occupied by a collection of elements is a *state density diagram* (Figure 4.25, bottom). The state density diagram relates how many particles are in different states. The values on the horizontal axis represent the state energy, and the values of the vertical axis represent how many particles are occupying the particular energy state. The state density diagram tells us in a glance what the distribution of particles in the system-of-systems is across the possible energy levels, but leaves out the degeneracy of energy levels.

As there are now many particles, it is possible to conceive of some particular organization of these particles in their possible states. Whenever there are multiple entities and multiple state possibilities, one can conceive of an organization (or lack of organization). This question of the degree of organization of the system-of-systems has been identified as a new property (an emergent property) called entropy.

Entropy is a measure of randomness. Entropy is all about how the stuff in a system-of-systems is organized with respect to the possible states (positional, energetic, and otherwise). This is represented in Figure 4.26. If there is little spread in the states that are occupied by

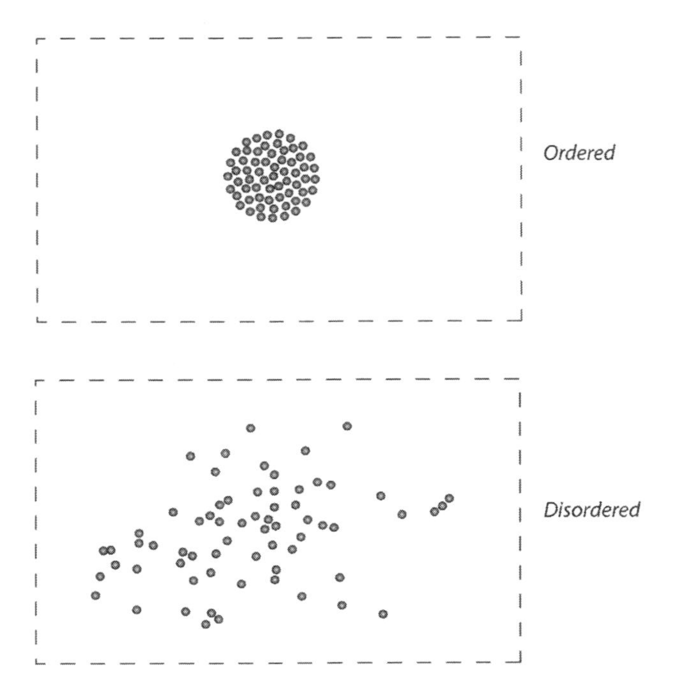


*Figure 4.26*

systems in the collective, then the entropy is low because it is necessarily known how and where everything is. We say that there is order. If however, there are a wide variety of states that the systems in the collective occupy, then the entropy is high, because it is less clear where or how everything is and how it will change. In this case, we say that there is less order in the system-of-systems.

There is a universal regularity (called the second law of thermodynamics) stating that if a system-of-systems is left unattended by coherent energy inputs, it will necessarily and spontaneously proceed in a way that leaves the system-of-systems with more entropy, and therefore less order. A gas spontaneously expands to fill a volume. A cup drops and spontaneously fractures into many pieces. These scenarios favourably increase the entropy of the universe. The opposite — the pieces spontaneously reassembling into the original cup — will rarely (never) be observed. If a box of toothpicks is dropped on the floor, they will spontaneously scatter in a pile, pointing in random directions, due to the absence of an organizing energy influence.

Our everyday intuitive knowing of entropy is as the inevitable wearing down and wearing out of things. It seems this universal drive to maximize entropy is the reason that we have to work and toil, as 'useful' things do not happen by themselves, but require an energy input or a directed force to make them so. As our mothers have always told us, things do not happen by themselves. The environment does not respond to our needs or our wishes. We must invest energy to make it so. A bird must collect sticks and build a nest, a groundhog must dig a burrow, and a human must make, carry, stack, and cement bricks to build a house. The organization humans generate through toils is subject to the ever-present, non-discriminating universal drive to maximize entropy, so that even in the wake of our efforts, things wear down, wear out, dissolve back into individual systems with randomly mixed components (as shown in Figure 4.27).

The only *apparent* exceptions to this second law of thermodynamics seem to be living systems (see Figure 4.28). Living systems *seem* to violate this fundamental law of entropy as they *spontaneously* create intricate systems, such as an apple tree or a baby girl, usually from

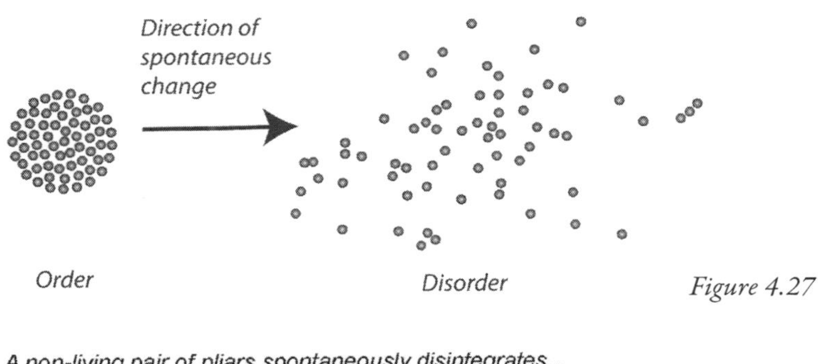


*Figure 4.27*

*Figure 4.28*

nothing much more than a single seed or cell and some random energy inputs and conditions. Living systems seem to effortlessly fight, or at least temporarily cheat, entropy as they spontaneously sprout, grow, heal, and resist disease and noise from a number of sources. While we still do not exactly know how living things accomplish what they do, we do have a grasp of the underlying impetus for their ever ordering behaviour, which is discussed in a following section (The spontaneous organization of the sun eaters, pages 137–142).

We have of course noticed that water (a collective system of water molecules), and other systems-of-systems like the magnetic spin systems discussed earlier, can spontaneous become more or less ordered as a parameter called *temperature* increases or decreases. Temperature is related to how much *heat* the substance has. The system-of-systems assume fundamentally different organizations (solid, liquid and gas) as the temperature changes and heat is absorbed or emitted from the substance. Let's examine some of the basic concepts behind temperature, heat, and phase changes, as these ideas are central to understanding the properties and spontaneous processes of living collectives, presented in Chapter 6.

## *Temperature and heat*

Temperature and *heat* are two things that we have all had experience with. On an intuitive level, heat seems to be almost like a fluid that flows out of hot objects and into cold objects. We associate temperature with a heat source, and we see temperature as the quantification of how much heat there is in the object. We find that if two things at different temperatures are put together, the heat *flows* from the hot body to the cold body so that they come to the same intermediate temperature. We also know that objects at temperatures higher than the environment will spontaneously 'release heat' to the environment, and that it takes energy input to heat them back up again. We know that if we take a block of ice and 'add heat', that there is a certain critical temperature (the melting point) at which it will change to become water. At this first level of comprehension, we know heat as a sort of fluid, and temperature as the measure of how much heat is *in* an object.

Let's increase our conceptual zoom to think of the molecules of an object — the systems of the system-of-systems. At this different perspective another completely different comprehension of heat and temperature is possible. In this context, heat and temperature are related to the way kinetic energy is distributed amongst the elemental stuff of the system-of-systems. Refer to Figure 4.29 and Figure 4.30 to assist in comprehending the reductionist view of temperature that's discussed in the next few paragraphs.[4]

At this new perspective, temperature represents the amount of possibility associated with the distribution of particles amongst their kinetic energy states. Heat input represents an energy exchange that increases the dispersion of the little system elements in their energy states. This corresponds to an increase in the temperature of the system-of-systems. This means that the entropy of the system also increases with increasing temperature. Temperature is related to a component of entropy specific to the distribution of particles in kinetic energy states.

At low temperatures, the probability of finding the system at low energies is very high, and therefore, because we know where and how the particles are (in the lowest kinetic energy states, moving very slowly), it is ordered and the entropy is very low. From this perspective, at absolute zero temperature, all of the particles of a quantum system are in the lowest ground states, and this constitutes perfect order. Consider the particle configuration in Figure 4.29 and Figure 4.30 to get an idea of what this means.

If energy is added to a system at low temperature, there is a large change in the entropy as many higher energy state possibilities become

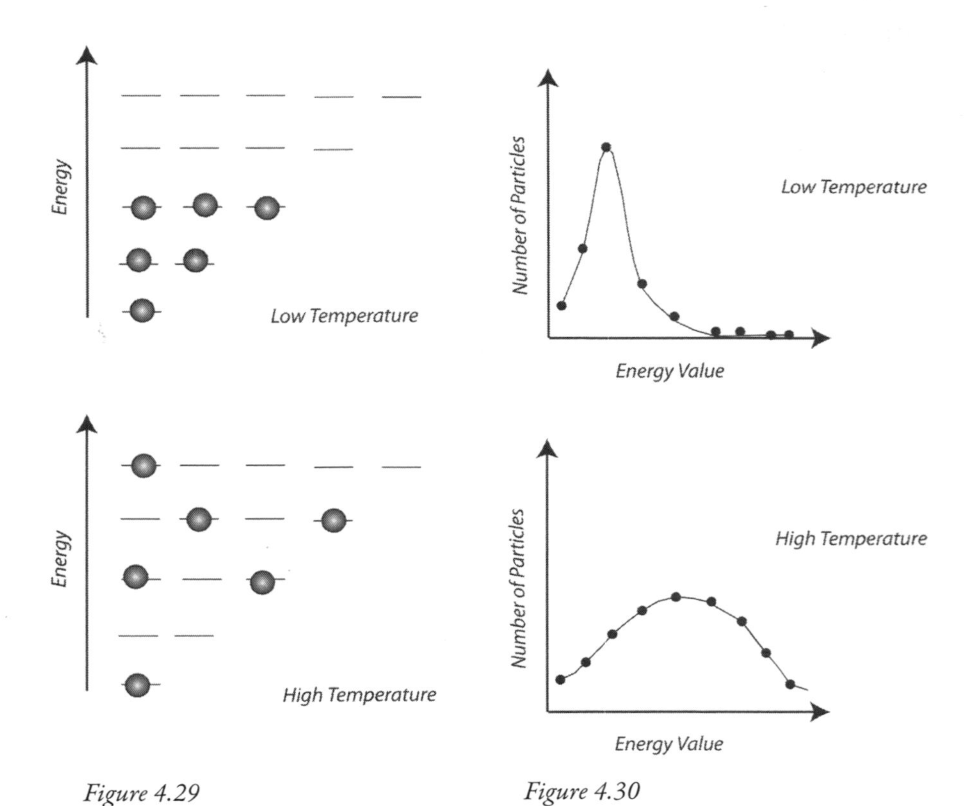


Figure 4.29 Figure 4.30

available to the particles. At very high temperatures the probability for the particles to take on any energy state is equal, and no states are favoured over others, even ones with very high energy — there is randomness, disorder and higher entropy. Compare the top and bottom images of Figure 4.29 and Figure 4.30 to get an idea of what kind of change this represents. Instead of all of the particles becoming concentrated at higher energy states at higher temperatures, they spread out more evenly throughout all possible energy states.

## *The spontaneous organization of the sun eaters*

In contrast to the behaviour of most non-living things, the spontaneous, elaborate and nearly perpetual self-ordering of living things appears quite perplexing. In fact as we have discussed, science has identified a law of the universe (the second law of thermodynamics) that notes spontaneous processes to occur only if they maximize entropy (disorganization). Some have proposed that living things can manage a local decrease in entropy by increasing the entropy of their surroundings. Considering more advanced life forms such as mammals, we can see how this is the case. Mammals spontaneously self-organize, but are consistently releasing heat (a highly entropic energy) to their environments. However, this release of significant amounts of heat (disorder) in return for organization and order is much less apparent, and possibly even non-existent for systems such as plants and reptiles which are in thermal equilibrium with their surroundings. This may lead one to believe that living things somehow manage to at least temporarily cheat this law of the universe.

Yet the second law of thermodynamics, as it's conventionally stated, was derived by studying *isolated* and *closed* systems. Isolated systems are unable to exchange anything with their environments. Once set up, an isolated system remains simply that — isolated. A closed system is able to exchange energy, but not mass, with its environment. Both isolated and closed systems are able to reach something called *thermodynamic equilibrium*, a state where all thermodynamic variables such as temperature, pressure, density and concentration become thoroughly mixed

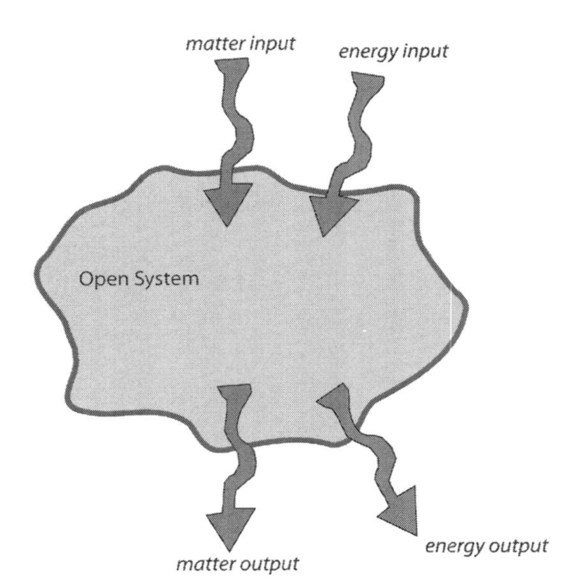


*Figure 4.31*

(homogenized) throughout the entire system. As the word 'therm' is often used in reference to temperature (thermals, thermometer, and so on) people sometimes mistakenly think of thermodynamic equilibrium as referring to equilibrium of temperature. Keep in mind that it refers to all kinds of system-of-system properties, and that the inevitable dispersal of a drop of ink in a glass of water is a good image of a closed system coming to thermodynamic equilibrium. A key feature of closed systems is that due to their spontaneous progress towards an inevitable homogenized state, isolated and closed systems can ultimately not store a pattern for any length of time. Moreover, isolated and closed systems are particularly artificial abstractions, and cannot be readily found in nature. They can however be readily found as human constructions in the scientific laboratory or machine design shop. The foundation of our conventional thermodynamic sciences is based on observations of isolated and closed systems.

Living things are much better represented by an abstraction called an *open system* (see Figure 4.31). An open system can exchange both matter and energy with its environment. A section of river flowing from a glacial mountain-top, over rocks, and to the sea is an open system. An uncovered pot of water heated on a stove receiving heat from below and loosing streams of water vapour from above, is an open system. All living things are open systems, constantly absorbing and secreting streams of energy and mass.

Plants are the sun eaters. They continuously absorb light energy from the sun, and streams of matter such as carbon dioxide, water, phosphorus and nitrogen from the atmosphere and soil. In turn, they continuously emit streams of matter such as oxygen, water, and various biologically assembled molecules. Unlike closed systems, open systems do not reach a thermodynamic equilibrium. Instead, open systems can operate in steady states of behaviour, but these steady states are characterized by various patterns of *inhomogeneity* of their inherent properties. These steady states may be the flow paths of a river over rocks, the form of bubbles in a boiling liquid, and the forms and associated functions of living things. Furthermore, a system may be dissipative, which means that it has a tendency to continuously release energy or mass to its surroundings. The second law of thermodynamics, as it is phrased for closed systems, does not apply to dissipative open systems. To get an appreciation for why open systems spontaneously assume intricate long-range patterns requires a consideration of energy quality. The quality of energy relates its ability to do something useful in the world (work) such as turn a windmill or fire a piston in an engine. Energy quality is described by the term *exergy*. More exergy relates a higher energy quality.[5]

Imagine river water at the top of a hill that flows downhill, picks up speed, but then slows down and collects in a pond. In this process the total amount of energy remains the same, but it is ultimately converted from high exergy potential energy, to low exergy heat energy. The potential energy of the water at the top of the hill due to the Earth's gravitational field is first transformed into kinetic energy (measured by the water's flow velocity) and heat energy (measured by the water's temperature). As the water continues to flow, to slow down, and pool into a pond, the kinetic energy it obtained from flowing downhill is eventually all converted into heat energy. Water at the hill top is thought to be of high quality (high exergy) because it can be used to do something useful, such as to turn a wheel and mill rice. The slightly warmer water in the pond is of lower quality (low exergy) as it's not as capable of exchanging its energy to another purpose (what work can we do with slightly lukewarm water?). Exergy has been destroyed as the system has come to equilibrium. Since high exergy energy relates to low entropy, and vice versa, low exergy energy relates to high entropy, entropy has been created in the process of the system coming to equilibrium, and this of course is favourable to the second law of thermodynamics, which likes to see the entropy of the universe increasing in a spontaneous process.

A restatement of the second law of thermodynamics for open systems comes in terms of exergy. It turns out that dissipative open systems will spontaneously self-organize, showing order over large length scales, if this organization gives them an enhanced capacity to degrade exergy. In actively adapting to better destroy the exergy of an energy source, high quality energy is converted to low quality energy, and thus, entropy increases. The non-intuitive aspect of the situation is that often highly collaborative, long-range, intricate organizations of the open system may arise because these kinds of structures are often more efficient exergy destroyers than randomly moving individual particles. The entropy decrease resulting from long range organization is much lower than the favourable entropy increase of degrading exergy. Thus, the adaptive open system may assume an intricate organization in order to better degrade exergy, create entropy, and assist in moving itself to a stable equilibrium state.

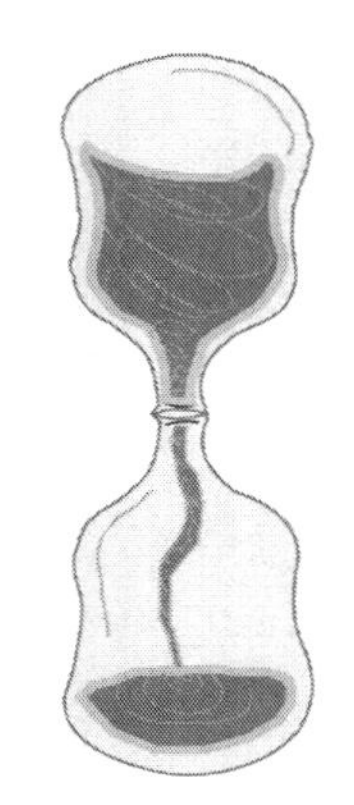

*Figure 4.32*

A child's toy called a *tornado in a bottle* exemplifies these concepts (see Figure 4.32).[6] This device consists of two pop bottles fused together at their mouths with some water inside them. When the toy is turned upside down, the water comes to possess gravitational potential energy and begins to flow into the bottom bottle. This represents a non-equilibrium state, which the system is trying to bring to equilibrium in moving the water to the bottom bottle. If however a slight rotational movement is given to the toy while in the non-equilibrium state, a spontaneous self-assembly of the system occurs, and a coherent water vortex (tornado) is established in the bottle. Trillions upon trillions of water molecules will suddenly be acting coherently. The purpose of this self-assembling tornado is to destroy the high exergy potential energy as quickly as possible. When in the tornado state, the water in the top bottle drains to the bottom in about 15 seconds. When in the incoherent flow state, the bottle drains in over 2 minutes. In this example we can note that self assembly happens in order to destroy exergy most efficiently, which takes the system to equilibrium. Favourably, entropy is created in the process.

Now, how does this help us to understand the spontaneous self-organization of the sun eaters (plants) and other life forms? Sunlight just so happens to be an incredibly high quality (high exergy) energy source. Moreover, sunlight impinging on planet Earth warms it up, which takes Earth farther away from an equilibrium state. Plants can be seen as dissipative open systems that have adopted intricate organization in order to destroy the exergy gradient imposed by the high quality energy of sunlight impinging on planet Earth. Plants absorb high exergy sunlight, and convert it into a much lower exergy chemical energy such as sugar and other biological molecules. In turn, plants are consumed by insects and animals and this chemical potential energy is converted to still lower exergy forms. Thus, the immediate *thermodynamic* purpose of life on Earth is to serve as an exergy destroying agent. Life continuously adapts and becomes more intricate and organized in order to better destroy the sun's exergy. Nested chains of living systems form with each feeding off the next, in a process of gradually converting more and more of the sun's high quality energy to a low quality, high entropy format.

This understanding of the spontaneous self-organization of the sun eaters in terms of exergy degradation is helpful to see how and why self-organization occurs and show that it occurs to fulfil the expectations of the second law of thermodynamics. However, what isn't explained is why living things are subject to a life cycle. Individual organisms begin as a seed and grow, expanding and increasing in intricacy and holding a manifest pattern with additional temporal activities, but then that mature pattern eventually begins to blur, blend, sag, brown and age. Inevitably, the organism dies and its form decomposes back into homogeneously mixed elements with its surroundings. Even collectives of organisms such as an ecosystems follow a life cycle in progressing through a well recognized process called natural succession. A barren plot of land will begin to grow grasses and other simple plants, and then small bushes and shrubs, and move on to become a forest of pine, and eventually of deciduous trees. An explanation of the life-cycle of organisms is not evident in the restatement of the second law of entropy for dissipative open systems discussed above. Nor is a parameterization and description of the different phases which occur during the life cycle.

A simple account of these spontaneous processes with the ability to predict and influence them would be of great benefit to ecology, and to the biology of whole organisms. Perhaps there is an expansion of the morphogenetic field concept which would bring us to this point? Get ready, as we will expand the field concept in Chapter 5, and will move on to account for the spontaneous processes of living things in Chapter 6.

## *Energy!*

Throughout our walk in the mind garden of physics we have continuously been referring to *energy*. Energy has been associated with the movement of particles, with the undulation of stuff-based waves, with the pattern-state of stuffless waves, with the ephemeral structure of electric and magnetic fields. Energy has been that unknowable something underlying the various happenings in the universe. In essence, physics is the study of the perplexing dance between matter and energy. Each thought system of physics considers a different aspect of this multifaceted, primordial story. In studying the ways of moving particles and waves, physicists attempt to understand how energy makes matter move, spin, undulate, vibrate, or fly. In quantum mechanics, physicists work very close to the interface between matter and energy. The dance is very odd at this interface, and very difficult for our human minds, accustomed to the ways of our everyday realities, to make sense of. In thermodynamics physicists work to understand the rules of energy allocations in a collection of many smaller elements. In understanding self-assembly, the physicist has learned that the system adopts this more intricate organization if it helps it to degrade high quality energy. All these things are ultimately about energy, and yet, we can scientifically know and work with energy only by accounting for its various measurable and demonstrable *effects* on the material things surrounding us.

Energy is the one thing in the universe that is immortal. Energy never dies, but rather, is transformed from one representation to another. It may start out as energy associated with the moving water of a river, be converted into the energy of motion of a paddle wheel in the river, and that paddle wheel may turn a generator which converts the energy of motion into an electrical manifestation of energy, and that electrical energy may flow through a piece of metal wire in a bulb to make light, and that light energy may ultimately be transferred to the contents of a room where they make the molecules in all the stuff jiggle just a little faster to end up as heat energy. Similarly, energy is never born. The energy that exists in the universe is all the energy that ever has, and ever will, exist. Albert Einstein showed us that matter itself is very concentrated form of energy. We have verified Einstein's idea by 'splitting atoms' to release some of the energy of matter to make electricity (nuclear reactors) and to wage tremendous mass-destruction in war (nuclear bombs). So we see that the whole universe, and all of the stuff within it, is really energy in one form or another.

We do not formally recognize life to be energy. We recognize that there is chemical energy in the bonds between the atoms that make up the living thing's physical body. We recognize that the living thing produces energy in the form of heat and motion, and that this heat and motion come from the chemical energy of the food that the living thing has ingested. But we do not recognize a distinct form of energy synonymous with the life of the organism. We once did, and still refer to our livingness as energy in casual language, but strangely, science no longer allows this pattern of thought in its story of reality. Even more oddly, there is nothing in science to prove that life cannot be another manifestation of the energy of the universe. It's just that for various historical and political reasons, the notion of life-as-energy has become an outcast from science, while other ideas and ways of perceiving and thinking about living things have settled and ossified in its place. Consequentially, as a culture we have largely given up the search for a way to substantiate, integrate and work with life as a property in its own right.

## *Towards understanding life and things living*

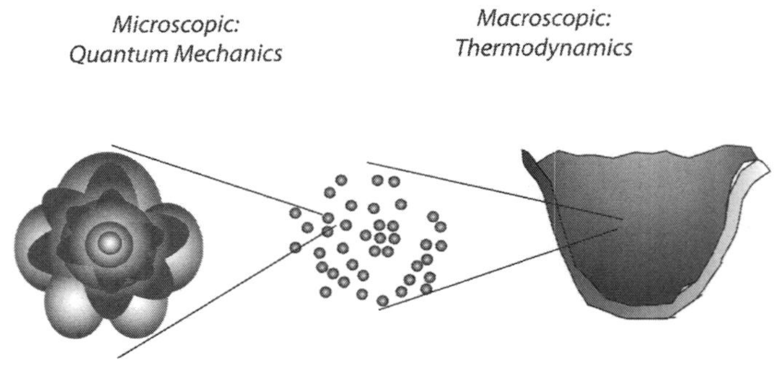


*Figure 4.33*

In our walk through the mind garden we have seen how human beings have developed conceptual systems capable of effectively structuring our experiences (observations) with the physical universe at different levels of perspective (shown schematically in Figure 4.33). In science, these perspectives are currently foliated as the *microscopic* realm of atoms, electrons and nuclei, the *mesoscopic* realm defined by primary organizations of these elements, and the *macroscopic scale,* which has relevance to everyday human proportions.[7] Quantum mechanics is designed to structure observations of the microscopic, individuated elements of matter. These microscopic elements behave much differently than matter we encounter in macroscopic systems, yet ironically, the stuff in our everyday worlds is composed of millions of trillions of these small elementary parts.

Thermodynamics structures our experiences with material processes at the macroscopic scale defined by human form. Thermodynamics, in dealing with enormous collectives of elementary microscopic parts, does not focus on what each individual element of the system is doing, but instead identifies new structures, properties and processes of the macroscopic system. Phases of matter such as the solid, liquid or gas are identified, and these refer to different interactions between and organizations of the microscopic parts. The macroscopic system has properties such as temperature, entropy, and heat, and phase changes occur in response to changes in these macroscopic system properties.

Thus, we can generalize and note that quantum mechanics is used to describe individual microscopic systems, while thermodynamics describes properties of macroscopic systems-of-systems. The middle scale (mesoscopic) is handled by statistical mechanics, which has incidentally bloomed into the exciting new science of systems and complexity theory. Systems and complexity theory aims to explain emergent properties in terms of the interactions and relationships between the reducible material parts.

The scope of this book is a little different than all of these perspectives, but makes use of concepts from each one of them. In Chapters 5 and 6, we will see how some of the unique aspects of living things can be structured using conceptual metaphors based on model ideas from quantum mechanics and thermodynamics. Rather than trying to describe emergent properties in terms of their parts, we are trying to work with emergent phenomena in a larger context, to characterize aspects of the super scale (uberscopic!) system to which living things may belong.

We'll call the use of quantum mechanics' ideas as modeling pictures for aspects of individual biological forms *organic mechanics*. As individual elements such as electrons and atoms come together into macroscopic system-of-systems for which new macroscopic properties and processes are identified, individual cells come together to form organisms and individual organisms come together to form a society or ecosystem. Amazingly, we'll see that many of the organizations and processes of collective life forms (such as the life cycle and natural succession) can be understood in terms of similar thermodynamic laws that we've already found to hold for non-living material systems. This understanding of collectives of individual living things in terms of thermodynamics will be referred to as *emergent thermodynamics*.

When systems of organic individuals are treated like everyday inorganic or non-living independent objects such as rocks or sand, unexpected, often counterproductive results may occur. Under the context supplied by organic mechanics and emergent thermodynamics, this is explained as the individual organisms and communities exist in distributions of previously unrecognized energy states which when disturbed, strive to meet equilibrium. Many of the ecological crises associated with human activities may be accounted for using the conceptual framework of organic mechanics and emergent thermodynamics. Organic mechanics and thermodynamics may engender new conceptualizations of living things, resulting in more effective understanding, interaction and integration of human activities with the natural world of which we are an integral part. These final aspects are discussed in Chapter 7.

# 5. Organic Mechanics

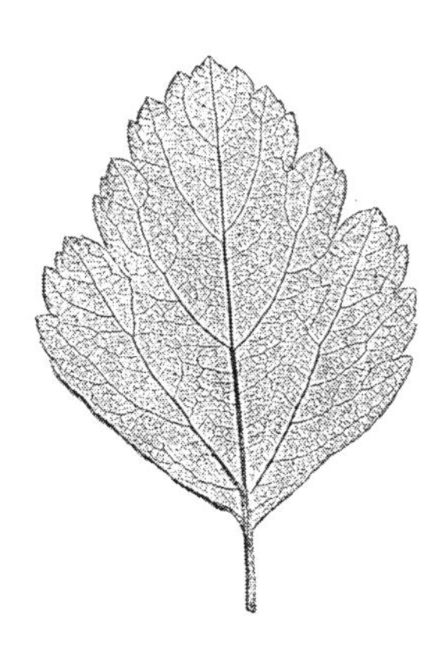

*Figure 5.1*

Have you ever looked at a leaf? I mean, have you ever looked at a leaf with the eyes of a child, and the eyes of someone who knows everything that you know right now, standing there together?

Leaves are exquisite things, so ordinary and ubiquitous that perhaps we've passed by how exquisite they are, but think on this: each leaf is an intricate pattern of relatively precise long-range order resulting from the replication and activity of a single cell. It is the study of these exquisite patterns that will allow us to exercise our imaginative rationality, drawing upon the mind garden to prepare deliciously synergetic new understandings.

## *The story of organic mechanics*

Truth be told, I could, and sometimes still can be, counted amongst those cynical, clinical scientists. Trained in the methodologies of the three -isms, I assumed the role of the objective, emotionally-distanced scientist without much questioning. In that role, a primordial sense of awe and wonder for the things of the natural world was somehow necessarily and strongly subdued. I went about with the (mistaken) assumption that if science could know the intimate workings of miniscule quantum particles and impossibly far away black holes, it had certainly elucidated all of nature's mysteries. Yet one particularly ordinary day as I sat eating my regular lunch on the outdoor patio of my favourite varsity eatery, a very subtle but poignant shift of perspective began when I noticed the leaves on the shrub growing next to my table (the same one shown in Figure 5.1). I plucked one off to examine what I felt to be a sublimely intricate pattern.

At that particular point in time, I knew that a leaf was composed of many independently acting cellular units, which had housed themselves within a structural matrix. I knew that many different types of cells made up the leaf, each type existing in a particular form (*differentiation state*), performing different roles as they took part in the different tissue systems. It was clear that the pattern that could be seen in the leaf resulted from contrast between two different systems: the vasculature, which like a highway system transported cell products and nutrients to and from the leaf; and the spaces in between the vasculature (*parenchymal tissues*), where existed cells with a primary role of transforming the energy of sunlight into sugars, for later use as food for the plant. As the vascular tissue cells were organized into tubes with a rigid structure, they appeared more raised than the cells involved in energy production. The energy-producing cells contained more of the pigment, *chlorophyll*, making them appear a deeper green than the veins. There was therefore a contrast of colour and form between the two types of tissue, which was the final pattern I observed in the leaf. I also knew that this leaf, like other biological forms, had developed this intricate pattern with a humble beginning as a single cell, or a small collection of initially identical cells. Therefore, I saw the mature leaf as a collective of hundreds of millions of cells that were about 10,000 times smaller than the self-created final pattern they ultimately found themselves a part of. This leaf, and living things all around, suddenly amazed me.

In the pattern of this leaf's veins I immediately noted multiple levels of structure, and some repeated pattern between these different levels, so that the leaf appeared to embody aspects of both wave and fractal geometry. I saw what appeared to be a broken wave-pattern, with the nearly circular wave-fronts (veins) appearing to emanate from the tip of the leaf, in a similar way that water waves appear when an object like a duck or a boat moves through still water, trailing wavefronts behind it. This broken 'wave-pattern' was repeated along each of the veins in the main pattern, and again along each of those finer veins. Eventually the space of the leaf became crowded, and the ends of the finest veins at the most intricate structural scale ran into one another to form a random array of closed polygonal shapes where no particular patterning was evident. Perhaps it was a bit strange to be fascinated with something as ordinary as this leaf from the bush where I had only moments ago sat eating a curried chicken salad sandwich and corn-chips, but that leaf caught a visceral level of my attention. At the time I didn't realize that no formal pattern was even recognized for leaves like these, and was even more surprised to find out that the developmental mechanisms for leaf vein patterns hadn't yet been sorted out. This pattern in the leaf remained a genuine mystery, even to modern day science.

As I walked home, the patterns of different leaves caught my attention. In blades of grass and the leaves of roadside tiger lilies, I noticed a simple pattern consisting of two endpoints where a set of gently curving vein paths met. This pattern reminded me of the fundamental harmonic of a regular wave (such as a wave on a plucked string) and due to its ubiquitous presence in the blades of grasses I called it the *blade pattern*. I then noticed the pattern in the leaves of apple trees, birch and catalpa, all of which embodied variations on the broken-wave structure of what appeared to be wavefronts generated and spreading out from the tip of the leaf. As the vein patterns in these types of leaves appeared to look like the skeleton of a fish, I called this the *fishbone pattern*. I saw in the leaves of maple and in wild grape a more complex pattern in which the fishbone pattern appeared to be scaled and reproduced at equal spacing around the point where the leaf stem joined with the rest of the plant. I saw these leaf patterns as the most complex of all, and called them the *rotator pattern*. Then, in passing a pine tree I noted the simplest 'leaf' in its needles, which didn't have much of a pattern at all, but was merely a tightly packed cylinder of veins. I called this the *needle pattern*. In looking around I saw these four pattern archetypes repeated in the leaves of every plant that I encountered or could think of. Cacti and pine trees used the needle

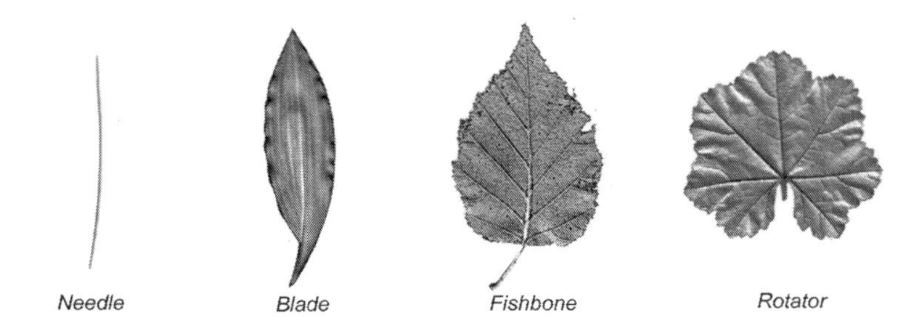


*Figure 5.2*

pattern; grass, palm trees, lilies, orchids and bamboo used the blade structure; oaks, dogwoods, catalpa, and birch trees used the fishbone structure; and maples, geraniums, wild grapes, zucchini plants, and water-lilies used the most complex rotator pattern. These were archetypal patterns, and their variations could be seen shared amongst all sorts of different plants (see Figure 5.2 for examples).

I took a sample of each leaf pattern archetype, and laid them all upon the kitchen table when I got home that evening. The patterns, needle, blade, fishbone and rotator, certainly progressed from simple to complex. Intuitively, I distinctly saw intricate pattern, possibly even pattern that could be mathematically defined. Moreover, I felt an inkling of purpose and meaning in those patterns, as they each seemed to be part of a larger system of wholes, analogous to standing wave harmonics or quantum wavestates that systematically progress from simple to complex (see Figure 5.3).

I stepped outside to the wide open view of the cedar deck and overlooked the mirror-calm stillness of the lake as the sun began to set. In what seemed like a fairly irrelevant train of thought, my quantum mechanics training suddenly flooded back into my mind. What if, I pondered, the interpretation of quantum mechanics that's so commonly taught in school (the so-called Copenhagen interpretation) was a bit off? What if there was no particle-wave duality, but rather, a particle like an electron was always a particle, but it was inexorably associated with a quantum wavestate as an ephemeral, overriding guiding pattern that influenced the particle's position so that we only ever measure the particle at some places and not others. In this interpretation of quantum mechanics (which is the essence of the pilot-wave interpretation of

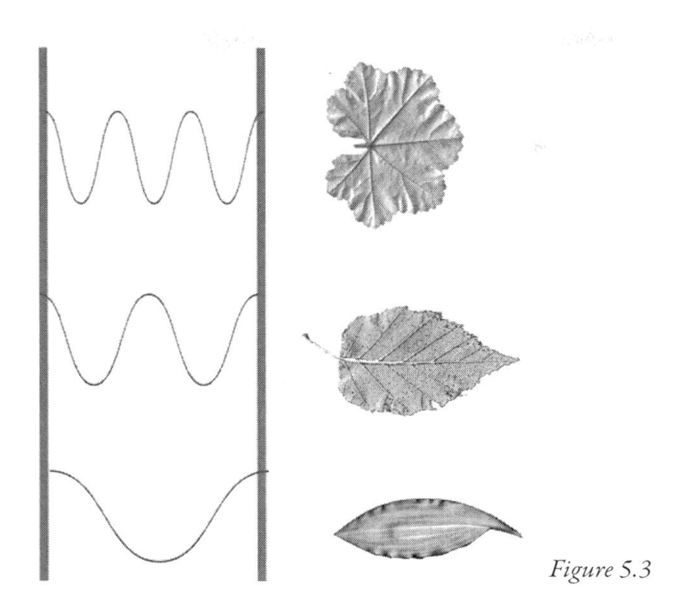

*Figure 5.3*

quantum mechanics developed by physicist David Bohm, see Chapter 4, pages 124–127), I imagined a collective of quantum particles guided by these ephemeral quantum mechanics waveforms into spatial patterns that emulated the pattern of their interfering, guiding waveforms. This alternative interpretation seemed to make a lot of sense, a lot more sense than the inconceivable paradox that something like the electron was both a particle and a wave at the same time.

I then thought of the leaves sitting on the kitchen table. What if, in some way, somehow, living things developed into a system that was not entirely unlike the system governing quantum particles? What if the matter (cells and molecules) composing these leaves were informed and directed into pattern by some sort of overriding influence, analogous to a quantum wavestate? Scientists had spent so much time tapping the extraordinary workings of impossibly small quantum particles to construct ideas that were far from intuitive and even seemed, to the uninitiated quite preposterous. What if, right under our noses, life forms were organized as systems that were not unlike those of quantum mechanics? After years of hard-core training in the three -isms style of thinking, I could appreciate how taboo ideas like these were in the scientific community. Yet, if it were possible that things at the atomic level should behave so extraordinarily, why should it be contentiously outrageous to propose a similar system for the intricate complexity of living things? The concept excited and compelled me to begin investigating it more deeply. I immediately called the idea of an association between biological forms and a long-range directing, patterning influence *Organic Mechanics*, and put aside my sceptical spectacles for just long enough to entertain the ideas to see where they might lead.

At the time, I didn't know about conceptual metaphors, vitalism, or the early embryologists and their morphogenetic field concept. It's when all of these ideas are put together that something of substance emerges with the potential to be forged into a scientifically respectable, practical format. In past chapters we've seen that conceptual metaphors are an inherent mental tool that structures more abstract, unfamiliar observations in terms of what we're already familiar with. They help us to develop new understanding and engineer new context in which to investigate and interpret a phenomenon. We've noted that conceptual metaphors are used in physics to structure basic concepts, and that this is a holistic, imaginative rationality that's already used in the sciences, but is entirely different than the thinking style of the three -isms. We've seen that the older, rejected or ignored ideas of the life sciences, such as life-as-energy and morphogenetic fields, are actually the seeds of befitting conceptual metaphors based on basic ideas in the physical sciences, such as energy and energy fields.

Now, if befitting conceptual metaphors are acceptable in physics, the foundation of science, then the notion of life-as-energy and morphogenetic fields should also be acceptable, or at least worth investigating, as appropriate conceptual metaphors for various aspects of life and things living. Moreover, we've noted that incongruent perspectives of a single phenomenon (such as the holistic and reductionist interpretations of heat) are simultaneously acceptable in physics, and therefore, it should be entirely acceptable that multiple interpretations of life and things living can also be held. A biological form can be seen as a reducible system of molecules that self-assembles structures to perform various functions and participate in various processes; as a complex, open system that can assume various dynamic, steady states of behaviour;

and as a whole supporting emergent phenomena which are capable of making things happen and are part of even greater systems.

Thus, in loosening the grasp of the three -isms perspective, it seems justifiable and intriguing to explore the notion of an influential field as a conceptual metaphor for biological development. At this point in time, morphogenetic fields have been almost entirely ignored and remain loosely defined in the sciences. There is no mathematical framework or description of what a morphogenetic field might be. Therefore, there's plenty of room to develop the conceptual metaphor, which is in part what this chapter is all about. However, even more intriguingly, it seems that the metaphor can be expanded to an entirely new level by making use of ideas from quantum mechanics as a new underlying model. In this sense morphogenetic fields become morphogenetic wavestates. This is promising as it opens up the context of a morphogenetic field as part of an even greater system of wholes.

In order to develop these conceptual metaphors involving fields and quantum mechanics, it's necessary to take the ideas at least a bit seriously, and to postulate the form of the fields/wavestates that may be (at least metaphorically) associated with biological forms, giving them some symbolic, mathematical representations. Although you'll see only the pictures produced from the maths, and not deal with the mathematical equations themselves, this is what we're going to do in the following sections. In order to do this, I choose leaves, whose patterns had fascinated me so much, as the first study system, and later, move on to include entire plant architectures in the theory. It turns out that while the theory of morphogenetic fields or morphogenetic wavestates is compelling, produces interesting images of the proposed fields/wavestates, and allows for the asking of new questions and therein opens new avenues of study; these are not the most significant outcomes. By extending the conceptual metaphor from one of fields to wavestates, the theory enters the domain of collectives, community, and ecology, where it can provide very simple explanations for the natural movements of whole systems (natural succession of plant communities, aging of an organism) and provide us with the opportunity to develop more harmonious interfaces and interactions with them. These further ideas are explored in Chapter 6.

## *Leaves: our chosen mascots for living things*

The leaf is going to be our mascot in the development of some ideas that will be considered extendable to other living things. There are a few good reasons for choosing leaves as a starting point. First of all, they manifest in intricate pattern with fairly precise long-range structure. They're also ubiquitous material of the everyday outdoor world, and are thus easy to find and collect without causing harm or suffering. Leaves from the same tree, and leaves from different trees that have grown under different environmental conditions such as variations in light, moisture, soil, and temperature, still show remarkable conservation of the leaf pattern (as seen in Figure 2.9, page 53). This insensitivity to obvious variations in environmental conditions indicates that the precise long-range order of the leaf is highly resistant to environmental noise, and is therefore tightly governed by the inherited algorithm of the plant. Moreover, the patterns of leaves are typically confined to two dimensions, so it's easy to represent and study them on a piece of paper, and to deal with them geometrically and mathematically.

## *How a leaf's vein pattern forms: a three -isms perspective*

What is known about the formation of vein tissue in plants? How is such intricate, long range pattern generated? Scientists have been working quite hard on these questions, but pattern formation in leaves remains an open problem.[1]

In a developing leaf, the vein tissue is known to arise in the earliest stages of development, at a point where the cells are all of the same type with no obvious differences between them. Scientists have worked out that flow paths of a substance called *auxin* are established in the growing cellular collective of the early leaf, and that these regions of high auxin flow then go on to become the veins.[2] The basis of this process is similar to the notion of water flowing down a mound of dirt. As the water begins to flow, the places of highest flow will move dirt along with them, which causes channels to form in the mound. In turn these channels can carry more water, which makes the channel deeper and even more effective at transporting greater amounts of water.

A similar thing is thought to occur in the formation of a leaf's veins. In the developing leaf, cells can self-produce auxin, but can't efficiently release this auxin to their environment. The cell requires another protein to actively pump auxin out to its surroundings; otherwise, the auxin is sealed within the cell. These auxin pumps are another self-manufactured item located in the cell's boundary. The interesting thing about these cell pumps is that it seems the greater the auxin flow across a particular cell-boundary region, the more auxin pumps the cell will place at that region. Just as a drop of dye in a bowl of water does not remain concentrated, but spontaneously spreads out to even out dye concentration throughout the bowl, auxin is also driven to spread to even out high concentrations developing in any region (as this spontaneous spreading out favourably increases the entropy). An interesting situation is therefore created in the developing leaf. The concentration of auxin is building up in cells which manufacture auxin, and wants to leave these cells to even out auxin concentration. However, auxin cannot leave the cells unless there are auxin pumps to actively move it from the cell. The number of auxin pumps a cell places at a particular location in their boundary is proportional to the amount of auxin flow at that point in the boundary, and the amount of auxin flow is proportional to the number of pumps already present. Thus, analogous to the case of water flowing down a dirt mound, the paths that initially support auxin flow go on to support more auxin flow, ultimately becoming an established flow path. Established auxin flow paths eventually become the veins of the leaf.

Scientists are only just beginning to understand this process, and cannot yet account for the relatively precise, intricate patterns of leaves using the above model. This is not to say that the model is incorrect, although it probably remains insufficiently detailed, and the actual situation more complex. It is therefore quite likely that the patterns in leaves have a reductionist, mechanistic, materialist explanation that does not need to bring in mystical or non-material based characteristics such as life-energy fields. However, let's assume that this situation of pattern formation in a living thing is akin to that of heat, which as we've discussed previously, has two entirely different interpretations — one of the three -isms and one of a more holistic nature — and both are useful. What would a holistic theory for leaf vein patterns that engages imaginative rationality be like? How would it be useful?

## *Morphogenetic fields for leaves: what might they look like?*

In the physical sciences, investigations into the nature of patterns that form in a material substance that are caused or related to energy fields have been an essential step in understanding the properties of the energies (and other factors) associated with those patterns. These patterns may be cracks forming under the influence of a mechanical stress field, an alignment of iron fillings under the influence of a magnetic field, or the interference pattern formed by diffracted quantum electrons. In other words, a first stage in the development of scientific understanding of any old phenomenon is typically the development of an effective mathematical description for the patterns associated with that phenomenon. The mathematical description is an important first step as it allows for a precise articulation and parameterization of the phenomenon. Explanatory theories involving further mathematical models may then follow. So we can consider the development of a mathematically-based description of morphogenetic fields for leaves as a necessary and important first step towards developing a scientific theory that more heartily encompasses morphogenetic fields.

The early embryologists of the 1900s to 1930s searched for the organizing principles behind biological form and employed the concept of a morphogenetic field, which like other fields in the physical sciences, was seen to denote cellular relationships and processes in the cellular collective, leading to specific biological structures being formed in specific regions of space. While no information has invalidated the concept of a morphogenetic field, it has been overridden by a strong impetus for explanations of biological systems in terms of molecular substances.[3] This is fine, and these reductionist models have proven quite useful; however, there are so many cells in an organism, and as each cell produces and interacts with an unbelievably large number of molecules in any number of different ways, comprehending the situation quickly becomes overwhelmingly difficult, even with the aid of today's computers. Therefore, a reductionalist perspective has limits that might be compensated for by considering alternative approaches. Morphogenetic fields may be one such alternative approach; not in that they assist in understanding the molecular scale interactions involved in leaf pattern formation, but that they allow us to move beyond reductionist perspective to consider the possibility of a larger system of wholes with its own importance. However, the idea of morphogenetic fields has not been taken seriously in the past decades, and consequentially, even a basic system of descriptive mathematics for morphogenetic fields does not exist.

Let's therefore re-think this idea of the morphogenetic field to determine if we can develop something more workable, and hopefully, more useful, in describing the properties and processes of living things.[4] Let's begin by simply *imagining* that from all of the complex activity in a cluster of cells, an entity analogous to an energy field can arise. This 'field', like other fields in the physical sciences, can be seen to arise from a source, to occupy space in a well-defined way (that is to say, it has a mathematical description), and to have a particular strength at different regions of space. Like other physical fields this field supplies a 'force' on an entity that is in its zone of influence. In this case, the entities are cells, and the 'force' acts to change the differentiation state of the cell. These fields are emergent properties, simultaneously depending on *and* dominating the behaviour of the cellular collective.

To appreciate how morphogenetic fields may help us to describe, understand and account for the patterns in leaves, let's consider the fishbone form of the Northern catalpa leaf (see Figure 5.4-A). Superimposed on top of the image of the leaf is one possible representation of a morphogenetic field that can account for the leaf's global pattern (Figure 5.4-B). This morphogenetic field is essentially a wave pattern with circular wavefronts that is being viewed from above. The white areas represent the high points of the wave, whereas areas of shadow represent the wave troughs. This morphogenetic field pattern is similar to the waves in water when a floating, moving object (duck or boat) is passing through. The high points of the morphogenetic field

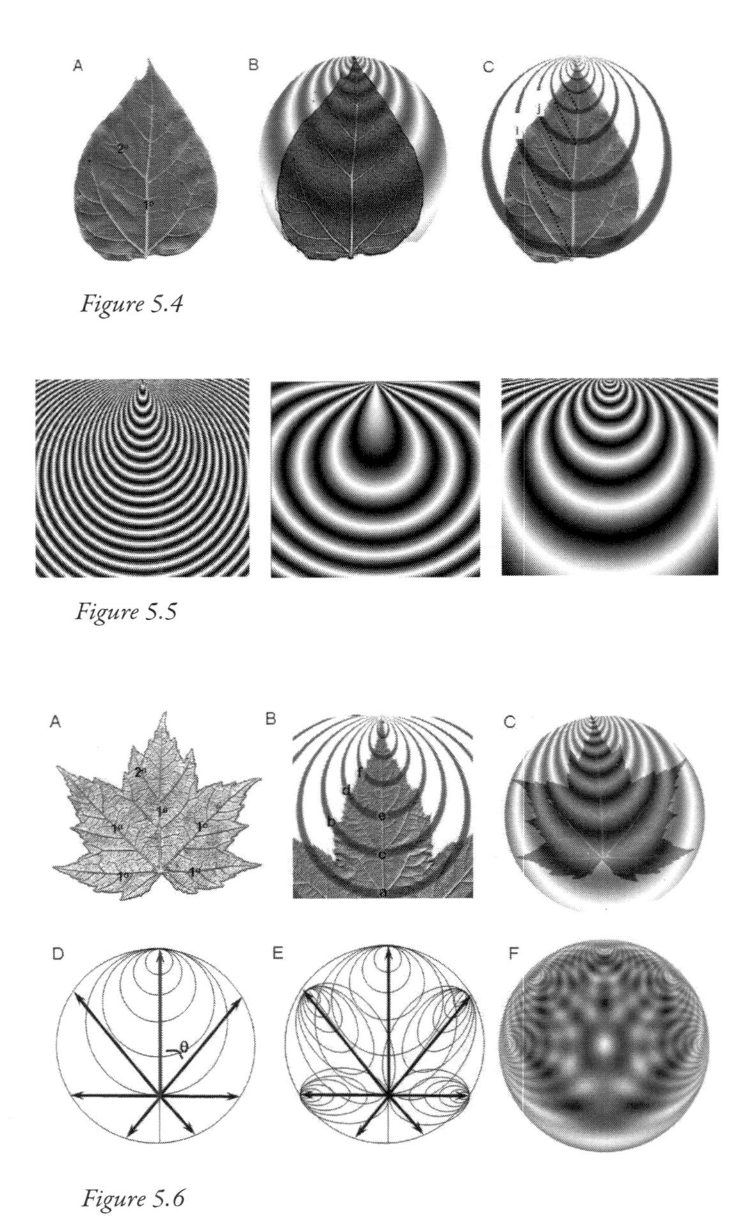


*Figure 5.4*

*Figure 5.5*

*Figure 5.6*

are what we imagine to have exerted an influence on the developing leaf structure (for clarity, only the high points are shown in Figure 5.4-C). Where the high points meet with the central axis of the leaf, the secondary veins originate from the midvein. Notice that the length of these veins is constrained by the next wavefront in the set. Thus, the spacing and length of the secondary veins of this catalpa leaf are directly related to the features of the morphogenetic field. This morphogenetic field has been generated using a mathematical equation. The 'source' of the field is located at the leaf's tip.

The morphogenetic field is generated by a mathematical function, and thus, has a precise description. The fishbone field has wave-like characteristics and is actually a 3-dimensional variation on the sine and cosine waves discussed earlier. The features of the field's wavefronts, such as their spacing, thickness and shape, are specified by parameters similar to the frequency and wavelength of sine and cosine waves. By adjusting the value of these parameters in the mathematical description of the fishbone field for instance, structural variations occur, and these can be seen to describe the variations in different leaves. Examples of these different field variations are shown in Figure 5.5.

For slightly more complex leaf patterns such as the rotator form of the maple leaf, there are more remarkable interrelations between the morphogenetic field of the central pattern and the length of the other primary veins in the set (see Figure 5.6-A). Just as in the case of the Northern catalpa leaf, the central patterns of leaves with the rotator archetype are described in relation to a wave-like morphogenetic field (Figure 5.6-B). If the mathematical equation generating the morphogenetic field describing the central pattern is used to generate one more wavefront in the morphogenetic field set, all of the primary veins fit within this new wavefront (Figure 5.6-C). This type of relationship is true for other types of leaves with the rotator pattern; the regularities are not just related to the maple leaf.

When different field variations are superimposed on top of their corresponding leaf patterns, some very striking relationships can often be observed between the form of the field's wavefronts and the form of the leaf's pattern. This provides even more evidence that morphogenetic fields are related to leaf patterns. Notice, for instance, the association between the wavefronts of the field and the secondary veins for the apple leaf (Figure 5.7). In the region near the base of the apple leaf where the wavefronts are the widest, there is a lack of precision in the location of the secondary veins, but as the wavefronts become narrower

*Figure 5.7*

*Figure 5.8*

towards the leaf tip, the location of secondary veins is specified much more precisely. For the dogwood leaf (Figure 5.8), note the close association between the shape of the wavefront and the shape of the secondary veins. In the case of dogwood, the field's wavefronts are very precise and the corresponding leaf's veins are also very precise.

## *Connecting morphogenetic fields and the three -isms perspective*

Now, to appreciate the actual features that morphogenetic fields may be related to in a developing leaf, and to develop an abstract tool for use in more general situations, let's consider the general beginning of a general organism or part of an organism such as a developing leaf. In the beginning, the cells of the seed, egg or developing leaf generally form a small, spherical or circular-disk shaped cluster of cells of the same type. The developing leaf is a particularly useful system for us to study as it begins as a thin circular disk of cells, so we can ignore its thickness and focus on the two dimensional circular face of the initial form. This is an easier geometry to keep in mind.

As mentioned in earlier portions of this book, through various biological mechanisms, localized regions of specialized cells commonly form as poles on this initial cluster of cells. Further changes in a developing collective can establish a central axis of specialized cells between these poles. Let's therefore imagine that the collective of cells representing a primordial leaf has developed two poles and a central

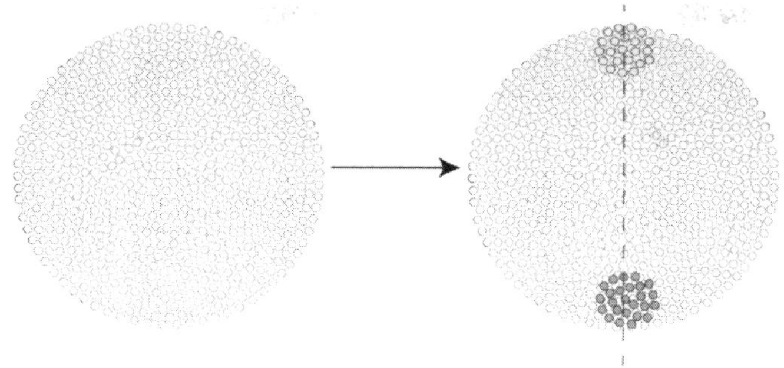

*Figure 5.9*

axis (Figure 5.9). For the specific case of the developing leaf, one of these poles may represent a group of cells producing auxin at what will become the leaf tip, and the other pole may be a group of cells breaking down or flushing out auxin at what will become the base of the leaf, where it joins with the rest of the plant. The central axis may represent the very first established path of auxin flow, which will later go on to become the midvein of the leaf. We will treat and think of these initial features (poles and central axis) as overall guiding features, similar to the first three nodes of the main triangle in the chaos game, which establish constraints and interconnectivity that allow the intricate, long range structure of the Sierpinski triangle to manifest (see Chapter 2, pages 59–63). The cluster, having broken the symmetry of the circle in forming these new features, is now in a position to develop more elaborate pattern.

Let's abstract (by simplifying and symbolically representing) this early, generic situation of a developing leaf to form a conceptual tool that we can work with. The situation will also be much more manageable if we can consider a case limited only to two-dimensions, which is why a study of leaf development is very interesting as they are spectacular patterns, but limited to two-dimensions. Imagine the circular, thin disk of the cellular collective is such a thin section that we can ignore its thickness and consider only the circular face. This face-on view allows us to best see the two poles and central axis of the developing collective. This abstraction is shown in Figure 5.10.

Let's consider this circular section with two poles and a central axis to be a quintessential *zone* of transformative activity. We shall consider

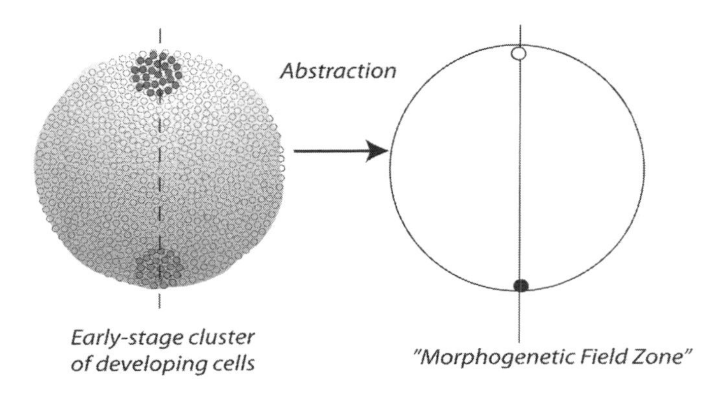


*Figure 5.10*

this zone to have the potential to give way to a variety of forms and structures, depending on the information patterns that are established within it. As we are moving away from details to consider the value in assuming a much wider perspective, we will not concern ourselves with exactly how this activity is generated, but rather, we will rest assured it *could* be explained by the networked activity of molecules, cells, and genetic information. These zones, in the expression of their inherent activity, develop into influential patterns that we call the morphogenetic field.

The generic features of a morphogenetic field for leaves, as discussed herein, are shown in Figure 5.11. We will define the pole at the leaf tip as the 'source' of the field's activity. This pole may be a literal source and control agent for an influential substance, such as auxin, which acts as a key substance in an algorithm carried out in the cellular collective. The circular boundary of the zone, which represents the initial edge of the cellular collective, acts to contain and constrain the cells and any chemicals and properties expressed within the collective, but gives way to change as the organism develops and new features form. Biological form need not fill the entire circular area, but is merely influenced within the area. The central axis of the field may serve as a specialized region of cells where specific responses to the pattern generated in the zone may be initialized. Therefore, we expect the features of the morphogenetic field, and the corresponding biological patterns it manifests, to have distinct geometric relation to the bounding circle, the poles, and central axis of the field zone.

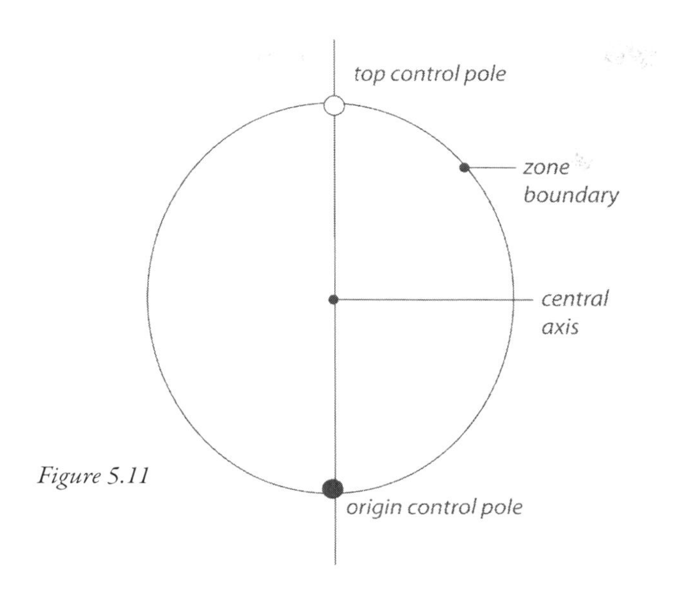


*Figure 5.11*

The morphogenetic field may represent something very real and physical, such as a pattern of morphogenic substance expressed in the cellular collective. On the other hand, while the morphogenetic field may have a resolvable physical basis, it also may not, and this does not necessarily discredit the 'existence' of the field. In this case, the morphogenetic field can be interpreted as an attractor for the non-linear processes involved in the developing cellular collective, analogous to the form of the Sierpinski Triangle when the Chaos Game is played. Consistently, other fields in the physical sciences, such as gravity, magnetism and electric fields did not, or still do not, have a readily conceivable physical basis. We observe the 'existence' of the field by virtue of the field's effects on the physical world that it influences. Morphogenetic fields can be thought of as one particular representation of an emergent property of the system.

Ultimately, morphogenetic fields are conceptual metaphors structuring observations of biological development that result from underlying algorithms in an interconnected, interacting collective of cells. Explicitly, the conceptual metaphor is one of a field. There are consistent parallels between the observations of biological development, and the concept of a field, and so, the conceptual metaphor can be recognized as a suitable fit. There may seem to be no inherent reason or benefit gained from assuming this perspective; however, this will become evident in further treatment of the concept, once we allow ourselves the chance to explore the idea and its implications in more detail.

## *A morphogenetic field system for leaves*

Let's suppose that for every leaf pattern, there is a morphogenetic field with an appropriate structure that emerges from complex processes inherent in the developing organism. This field guides the specialization of cells in space and time to generate a long-range pattern. Let's see if we can elaborate on the concept of morphogenetic fields to account for the vascular patterns of a great number of leaf types.

There are myriads of different leaf patterns in existence with some examples shown in Figure 5.12. However, the great diversity of patterns can be seen to stem from variations on four primary organizations, which we've called:

— the needle (Figure 5.13);
— the blade (Figure 5.14);
— the fishbone (Figure 5.15); and
— the rotator (Figure 5.16)

Therefore, four general types of morphogenetic field are required to account for each of these general leaf types. Variations in the parameters of each general type of field would be able to describe the form of all of the pattern variations with a particular pattern class. We will base all four basic types of morphogenetic field on the geometry of a circular zone with two poles and a central axis, as previously described and shown in Figure 5.11. Each field type will have inherently different activities, and construct inherently unique patterns within the circular morphogenetic field zone.

A description of each of the leaf types and its associated proposed morphogenetic field type follows below and are shown in Figure 5.17:

*Figure 5.12*

*Figure 5.13*

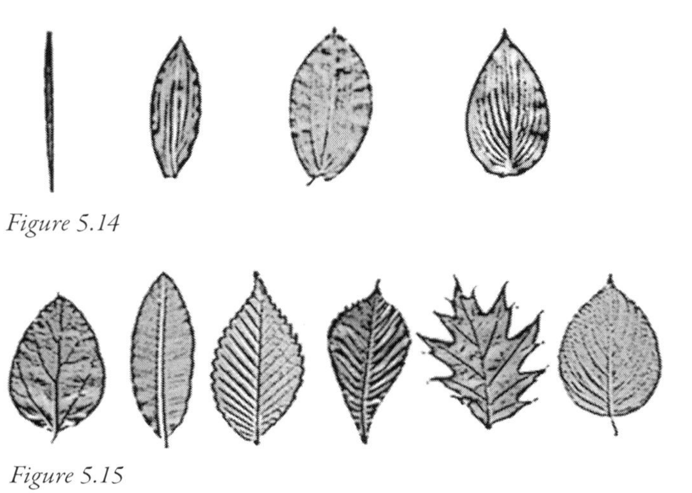

*Figure 5.14*

*Figure 5.15*

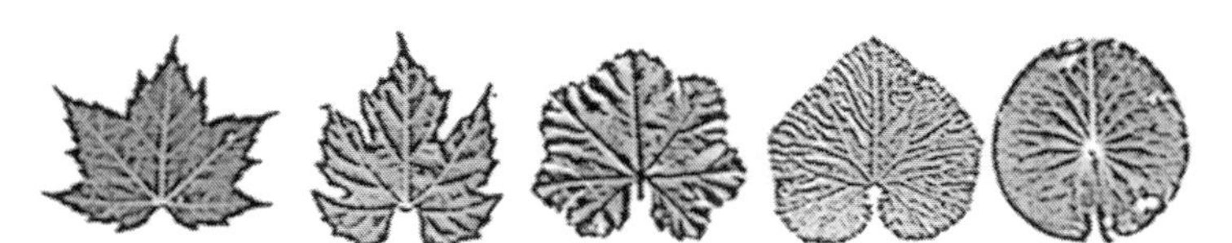

*Figure 5.16*

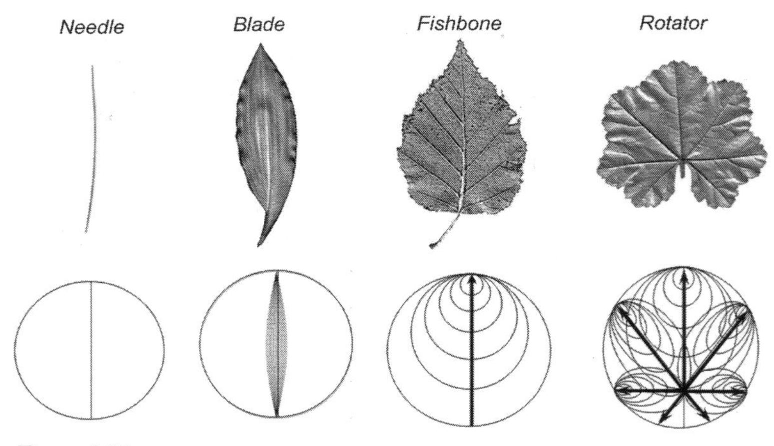


*Figure 5.17*

**Type 1**, the *needle* structure (examples: pine, cactus). This structure exhibits form only along a central axis. Therefore, the morphogenetic field can be seen to simply generate form along the central axis of the circular zone, connecting the two poles of the circle. The needle of white pine is shown in Figure 5.17, with the proposed corresponding morphogenetic field shown below the image. As only the radius, length, and a simple equation for the volume are required to describe this structure (3 parameters) this is proposed to be the simplest form of 'leaf'. Other leaves (or plant parts) which assume simple shapes without vein patterns (three- dimensional cylindrical arrangements, flat lobe-like leaves of some succulents) would also fall into this category.

**Type 2**, the *blade* structure (examples: grass, daffodil, palm). This structure exhibits a repeated, scaled, periodic pattern in the space defined by the two-pole field. Usually, the pattern is closed at both ends and is symmetric about the central axis. The bluebell leaf is shown as an example in Figure 5.17, with the proposed corresponding morphogenetic field structure shown below the leaf image. The blade structure is more complex than the needle structure as a minimum of four parameters are required to describe its structure.

**Type 3**, the *fishbone* structure (examples: leaves of roses, apples, oak, and chestnut). This form exhibits secondary veins that grow out at an angle from a primary, central midvein. The corresponding morphogenetic fields for this structure are circular waves with a shared point located at the top pole of the circular zone. The wavefronts of this morphogenetic field are imagined to directly influence tissue in their presence, or to guide the differentiation of tissue in a systematic manner dependent on the wavefronts and one of several possible 'guiding mechanisms'. Intersection of the central axis and the crests (maxima) of the circular wavefronts corresponds with the origin of secondary veins from the midvein. An example of a fishbone structured leaf is shown in Figure 5.17. Below the leaf image is an example of a fishbone field, where only the maxima of the field's wavefronts are shown as circular lines. The fishbone structure is proposed to be more complex than the blade type as a minimum of five parameters are required to define its structure.

**Type 4**, the *rotator* structure (examples: the leaves of maple, wild grape, geranium) in which the fishbone structure (or another structure such as needle or blade field) is rotated and scaled to form a broadened, web-like leaf. For leaf samples with a central pattern described by a fishbone structure, a wave-front containing the entire shape of the leaf (that is, all primary veins) can be obtained by extrapolating the morphogenetic field of the central pattern to include one more wave-cycle, which can be seen to constrain the length of all primary veins. Each of these primary veins then re-creates the morphogenetic field of the central pattern, scaled appropriately. Figure 5.17 shows an example of this leaf type, with an example of the associated morphogenetic field shown below the

> leaf image. The rotator structure with a fishbone structure central pattern is proposed to be the most complex of all two dimensional structures, as six parameters are required to define its features.

For each of these general field types, a number of variations are possible by changing the descriptive parameters of the field type. These variations lead to the great number of leaf pattern variations that we observe in the world today.

## *Morphogenetic fields for plant structure*

Let's see if the morphogenetic fields we've developed to work with patterns in leaves can be extended further to describe other patterns in plants, such as the placement of leaves on a stem. To see how this would work, let's consider a raspberry bush (as shown in Figure 5.18). Three levels of nested morphogenetic field are required to describe the plant's organization. A first morphogenetic field with a fishbone structure can be used to define the origin and the length scale for secondary branches from the main branch. The intersection of these secondary branches with the boundary of the circular wavefronts can be seen to call up secondary morphogenetic fields. These secondary fields with the rotator structure define five stems and five corresponding tertiary-level fields. These tertiary fields have the fishbone structure and act to pattern the leaves.

This example with the raspberry bush reveals that in trying to address plant structures, there are two dimensions to the problem. The first is the recognition of the number of levels of structural hierarchy existing in the plant (the raspberry has three) and the type of patterning that's present at a particular hierarchical level. These two features can be seen as the dimensions of complexity of the plant form. The greater the number of hierarchical levels, the more complex the plant (Figure 5.19). Yet, a greater intricacy of patterning at a particular hierarchical level also increases the complexity of the plant (Figure 5.20). On this basis, a maple tree is seen as more complex than a grapevine, as

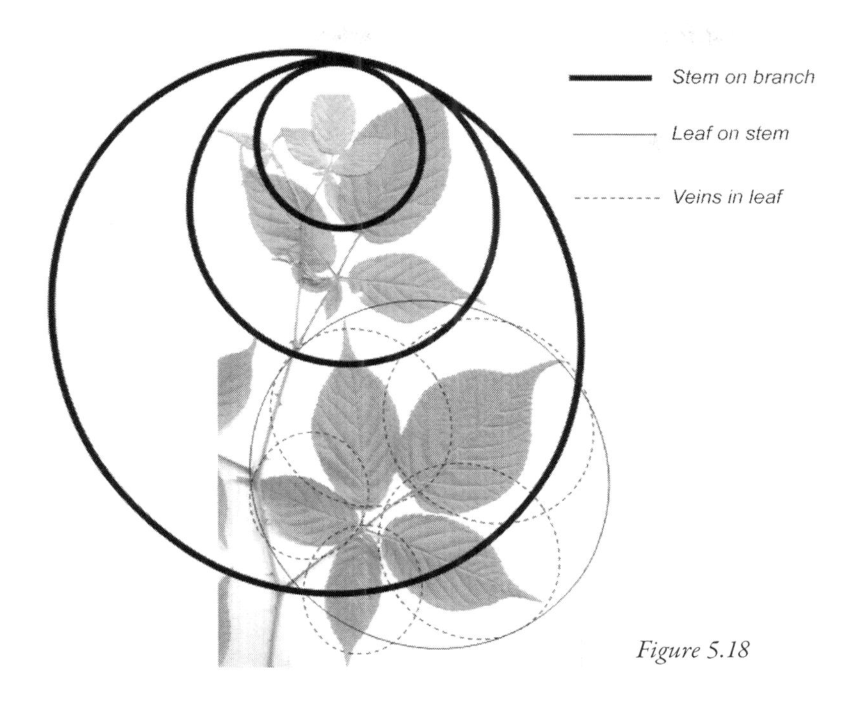


*Figure 5.18*

*Figure 5.19*

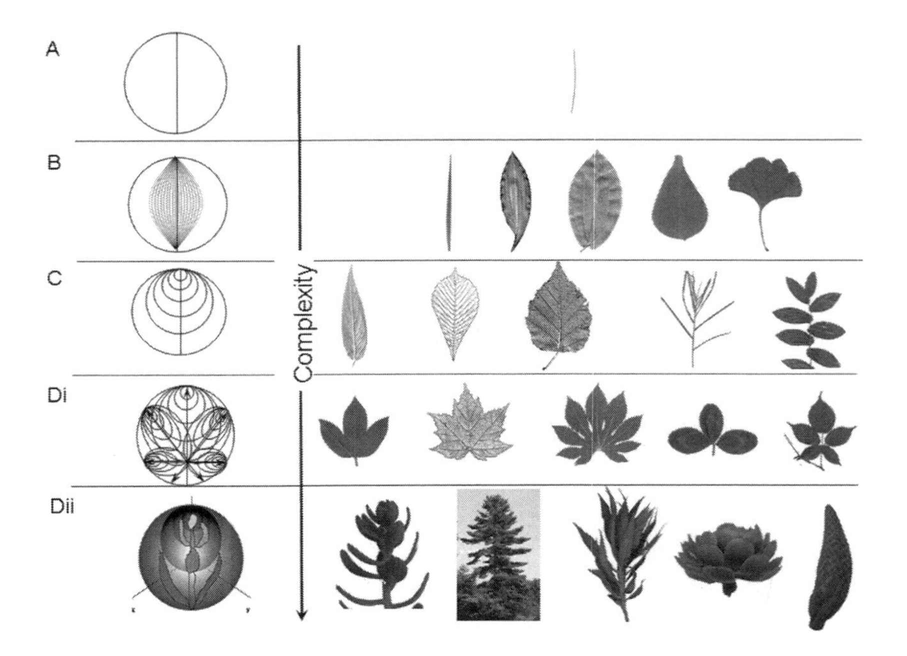


*Figure 5.20*

the maple has a greater number of hierarchical levels of structure, even though both plants share a similar leaf patterning. On the other hand, a maple tree is also more complex than a pine tree, even though both share the same number of structural hierarchical levels, as the maple tree has more complex leaves than the pine.

Thus, in working with plant architectures, hierarchy is important, and a general definition of the hierarchical structuring of different types of plant would be useful. Assuming a very coarse view of plant architecture, one which overlooks root systems and dynamic activities involving flowers, fruits, and seeds, one might say that the plant can be described by the organizational structure surrounding the leaves — that is the stems, branches and trunk of the plant. From the consideration of this coarse perspective, plants can be categorized according to the number of levels of structural hierarchy that they possess:

0. Algae — Algae are single cells with the ability to photosynthesize. Single cells may form collectives, but do not come together into an individuated, holonic body of form.

1. Mosses — Mosses are single, many-celled structural units lacking distinctly individuated systems such as distinct leaves, branches, or trunks, which are common to other plants.

2. Basic Plants — Basic plants have leaves with different venation patterns. They have a stem system on which leaves are organized. Common examples are grass, clover and milkweed.

3. Bushes — Bushes have a branch system on which stems are organized. Leaves are organized on stems. Leaves have characteristic venation patterns. The raspberry is a good example of a bush as there are clearly distinguishable patterns for the branch, stem and leaf systems.

4. Trees — Trees have a trunk system which organizes branches. Branches organize stems. Leaves are organized by the stems. Leaves have characteristic venation patterns. In trees such as maple, hickory and pine, there are clearly distinguishable arrangements of the different organizing systems of trunk, branches, stem and leaf.

Each of these organizing systems (veins of leaves, leaves on stems, stems on branches, branches on trunk) can exist as a clearly distinguishable level of hierarchy in the plant (see Figure 5.21 for a specific example). The previously described morphogenetic fields (needle, blade, fishbone, and rotator) can be used or adapted to describe the entire plant architecture at a particular hierarchical level. Note that the degree of precise, reproducible organization decreases dramatically in the higher levels of plant structural hierarchy, such as in branch and trunk systems.

*Figure 5.21*

In other words, the amount of noise from environmental influences, rather than genetic algorithms, increases in higher levels of the structural hierarchy. This can be accounted for by considering the time-scale at which features form. The veins in leaves form over the time scale of days, which is fast enough that processes of the biological algorithm can maintain an apt and dominant level of control over environmental influences. In contrast, the trunk system of a tree forms over the time scale of years. In such a long time span, it is more difficult for processes of the biological algorithm to maintain a primary influence. Plant parts such as trunks and branches are therefore most susceptible to environmental irregularities, and appear as more chaotic and randomized structures as a result.

Of course, unlike the vascular patterns of leaves, the patterns of plant architecture extend into three dimensions. Three-dimensional patterning in plants can be seen to take analogous forms to two-dimensional morphogenetic field structures, only instead of circular wavefronts, three-dimensional analogues have spherical wavefronts. For instance, the needle structure can be seen to assume the form of a full three-dimensional cylinder, and is of equivalent complexity. A very common three-dimensional pattern involving a spiral arrangement of leaves can be described by a three-dimensional analogue to the fishbone structure (examples are shown in Figure 5.22). The form consists of a nested set of spheres which are joined at the top pole point. As in the two-dimensional fishbone field, each sphere is seen as a wavefront with a particular influence on the structure. Leaves (or stems, branches, and

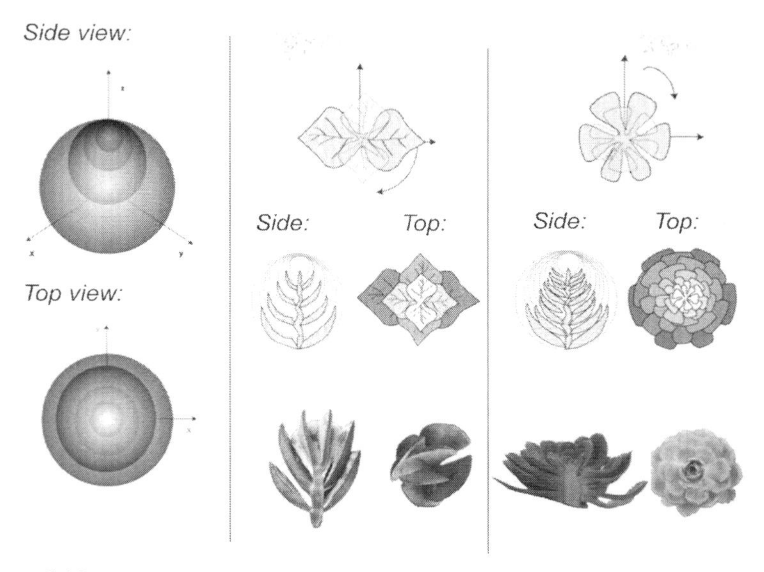


*Figure 5.22*

so on) originate at the point of intersection between the spherical wavefront maxima and the central axis, and grow outwards according to one of several guidance mechanisms involving the wavefronts. Each bud-point may exhibit a number of leaves. For each advancing wavefront and node in the set, the leaves are rotated about the central axis by a fixed angle. In lengthwise cross-section the spacing and leaf lengths are described and constrained by the morphogenetic field in similar ways to those of the fishbone structure. This uniquely three-dimensional structure will be called the spherical-fishbone structure and is depicted in Figure 5.22. The spherical-fishbone is of the same complexity as the rotator, as an equivalent number of parameters are required to define its structure.

## *The complexity number*

Intuitively, we observe that a tree has more structural complexity than a bush, a bush more than a basic plant, a basic plant more than a moss, and a moss more than algae. This is because each category indicates the existence of an additional hierarchical level of organization,

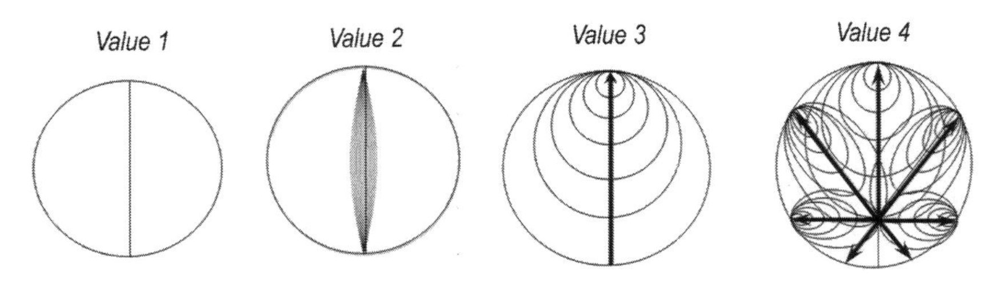


*Figure 5.23*

and therefore, entirely new systems with new organizations become possible. Within each of these categories there are also varying levels of possible complexity, attributable to the type of morphogenetic field involved in organizing the particular plant system.

Working in analogy to the quantum number, which systematically indexes the wavestates of a quantum wavefunction, it seems quite possible, and possibly quite useful, that a single number could be used to relate both the number of structural hierarchical levels and the pattern at each level, and therefore index the set of embedded morphogenetic fields involved in directing the form of a particular plant's architecture.

Our basic integer numbering system makes use of digits and specifies a value for each of those digits. Four systems of organization have been identified for a particular plant (these are the trunk, branches, stems and leaves) and there are (at least) five different morphogenetic fields involved in the structuring of each of these levels of organization (these are the needle, blade, fishbone, rotator, and spherical-fishbone fields). In developing a single number to specify a plant's complexity, we can let the number of hierarchical levels (plant category) specify how many digits the number will hold. On this basis, a tree, with four levels of organization, will be a four digit number, a bush a three digit number, a basic plant a two digit number, and a moss a one digit number. We can then assign a value to the morphogenetic field according to its inherent complexity (as shown in Figure 5.23). The needle structure, the simplest, is assigned a value of 1, the blade a value of 2, the fishbone a value of 3, and the rotator and spherical-helix structures a value of 4. We can let the value at the particular digit place correspond to the morphogenetic field used to organize that particular structural level. Let's call this resulting number the *complexity number* for the plant.

*Figure 5.24*

The raspberry bush has three levels of structural organization (stem on branch, leaf on stem and veins in leaves). Therefore, the raspberry bush is a three digit number, where the value of each digit corresponds to the value of the morphogenetic field used to organize the level of structural organization (see Figure 5.24). In this example, the largest digit refers to the stems-on-branches organization, which occurs with the fishbone field, and therefore has a value of 3. The next digit pertains to organization of leaves-on-stems, which is done using a rotator field, with a value of 4. The smallest digit, which pertains to the organization of leaf vasculature, is again a fishbone field, and has a value of 3. The complexity number for the raspberry bush is therefore: 343.

Under this system, a pine tree (refer back to Figure 5.21), which has branches organized using a three dimensional spherical-fishbone field (4), stems organized using a two dimensional fishbone field (3), leaves organized on stems using a blade-like field (2) and leaf structure of a needle field (1) will have a complexity number of 4321. Crabgrass, a simple plant with a fishbone field arrangement of leaves about the stem (3) and blade field organized veins (2) will be represented by the number 32 (see Figure 5.24).

The complexity number is useful as it allows us to specify the complexity and intricacy of any plant with a single parameter. Like the quantum number, the complexity number specifies a number of patterns states that characterize the entire form of the plant.

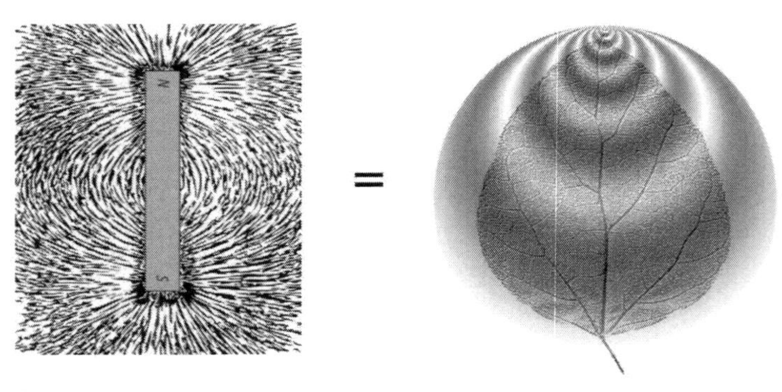

*Figure 5.25*

## *Morphogenetic wavestates*

On our journey through this book, we've ventured through a number of different conceptual vistas and are now ready to begin consolidating our experiences. Early on we saw that conceptual metaphors are a tool employed, most often subconsciously, by the human mind to cook up new understanding and new context for an unfamiliar phenomenon. We've seen how conceptual metaphors are used in the sciences, in particular the physical sciences, where modalities such as potential energy landscapes and heat as a fluidic substance are common examples of a metaphoric thinking style we've called imaginative rationality. Imaginative rationality is in contrast to the dissecting material literalism characterizing the thinking-style monopoly of reductionism, mechanism and materialism, which pervade the life sciences. We've considered that older notions, such as that of life-as-energy and of an influential field in biological development, have congruency with various pertinent observations of living things, and thus, can be considered befitting conceptual metaphors worthy of further investigation and development. After all, when we look objectively at physics many of the concepts, from energy fields to quantum wavefunctions, appear just as far-out and ungrounded in material reality as do those of life-as-energy and morphogenetic fields. So far this chapter has been primarily concerned with embracing the notion of morphogenetic fields as befitting conceptual metaphors and has pursued development of the idea using the patterns in plants as study subjects. The notion of the original conceptual metaphor we've begun with — that of a field associated with biological development — is represented by Figure 5.25.

In our walk through the mind garden, we've also examined the concept of a *stuffless wave* and the associated idea of a *discrete energy state*. We have considered life at the atomic level, where *wavefunctions* and their harmonic *wavestates* are associated with the position and energy of quantum 'particles'. We've seen that the wavestate is associated with the discrete energy state of the particle. For particular shapes of potential energy (which can most simply be known as environmental constraints), there are systematic manifestations of a set of discrete wavestates. For simple cases, such as a particle in a box-shaped potential, the wavestates manifest as spatial patterns akin to the harmonics of a standing wave (recall Figure 4.12, page 118). For different distributions of potential energy in space, such as a spherically shaped energy potential, the wavestates exist with hierarchical structuring akin to nested Russian babushka dolls, where higher energy values correspond to more levels of structural hierarchy in the set of associated wavestates (recall Figure 4.16 and Figure 4.17, pages 123 and 124).

The morphogenetic fields we've developed in this chapter, and the leaf patterns they're associated with, appear to belong to a larger, coherent system, just like individual harmonics of a waveform. It is this aspect that sparks the idea of extending our initial conceptual metaphor of a field associated with biological development to an entirely new and absolutely expanded level. This is imaginative rationality in action! David Bohm's pilot wave theory (discussed in Chapter 4, pages 124–127) interprets the quantum wavestate as a guiding pattern which specifies the probability of finding a quantum particle in a particular trajectory through space and time. Similarly, the morphogenetic field can be interpreted as a guiding pattern that specifies a probability of finding a cell in a particular place and state of specialization. As a guiding pattern, the quantum wavestate is also interpreted by the pilot wave theory to be akin to an energy field.[5] Similarly, the morphogenetic field has a number of field-like qualities, such as: a non-local dispersion through space, an originating source (here this is hypothesized to be at the tip of the leaf), a predictable effect on entities (cells) in its zone of influence, and a direct relationship between the geometry of the field and the influence of the field on form. Moreover, there are numerous parallels between the idea of a quantum wavestate and morphogenetic fields for cells. These parallels are summarized in Figure 5.26.

| Quantum Mechanics Wavestate | Morphogenetic Wavestate |
| --- | --- |
| Exists as positional information (of unknown origin) that is associated with, and acts to guide, the dynamics of a quantum particle in space and time. The amplitude of the wavestate is related to the probability of finding an electron at that location in space and time. | Exists as positional information (of unspecified origin) that is associated with, and acts to guide, the dynamics of cells in space and time. The amplitude of the wavestate is related to the probability of finding a cell in a particular differentiation state at that location in space and time. |
| The range of the wavestate's activity is much greater than the size of the quantum particle associated with it. | The range of the wavestate's activity is much greater than the size of the cells associated with it. |
| Manifests as a wave-like entity with classical periodicity. | Manifests as a wave-like entity with classical, scale, or logarithmic periodicity. |
| For spherical potential energies, the wavestates exist as geometric patterns and are hierarchically structured. | For the plant kingdom, the morphogenetic wavestates exist as geometric patterns and are hierarchically structured. |
| Represents the occupation of a discrete state of energy. | Represents the occupation of a discrete state of emergent energy. |
| The wavestate's pattern increases in complexity with increasing energy | The morphogenetic wavestate, and associated organic form, increases in complexity with increasing emergent energy |
| Quantum numbers index the system of wavestates and specify a particular state with a particular form. The magnitude of the quantum number is related to the amount of energy the quantum particle posseses. | Complexity numbers index the system of morphogenetic fields and specify a particular emergent state with a particular form. The magnitude of the complexity number is related to the amount of emergent energy the organism posseses. |

*Figure 5.26*

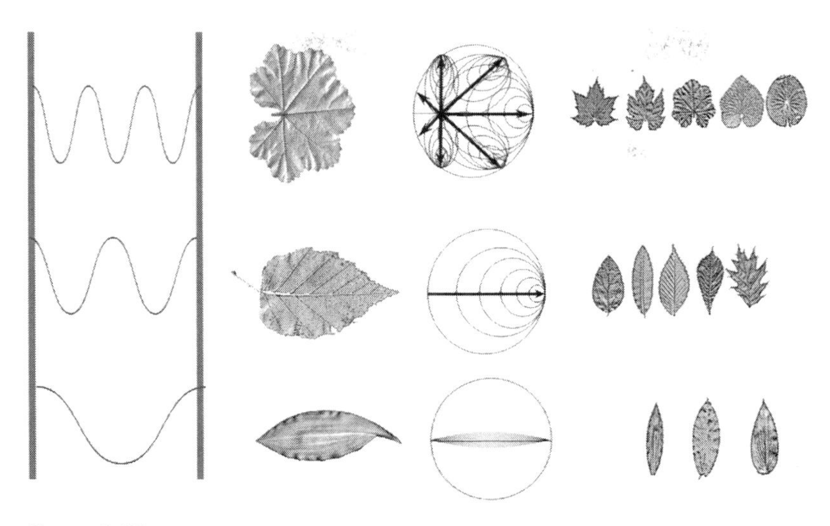

*Figure 5.27*

To wrap one's mind around the new idea, let's first only consider the patterns in leaves, and the morphogenetic fields that we've developed for them in this chapter. The archetypal fields (where again the archetypes are needle, blade, fishbone, rotator) systematically progress from simple to complex. While variations upon each archetypal theme occur, the basic form of each archetypal pattern remains conserved and is clearly identifiable. Each morphogenetic field is thought to be a long-range spatial influence directing cells into the appropriate place and differentiation state. This can all be seen to parallel the quantum particle-in-a-box system, where there is a systematic progression of standing wavestate harmonics from simple to complex, and the form of each harmonic remains distinctly conserved. Each quantum wavestate (harmonic) is a long-range spatial influence directing the quantum particle accordingly. For quantum mechanics this progression from simple to complex also corresponds to progression from low to high energy. Thus, each archetypal leaf pattern can be seen as a state in a set of 'morphogenetic wavefunctions' (as detailed in Figure 5.27). Moreover, we could follow suit and draw a complete analogy with quantum mechanics by deducing that these morphogenetic wavestates are also associated with an energy, and in progressing from simple to complex they are also progressing from low to high energy.

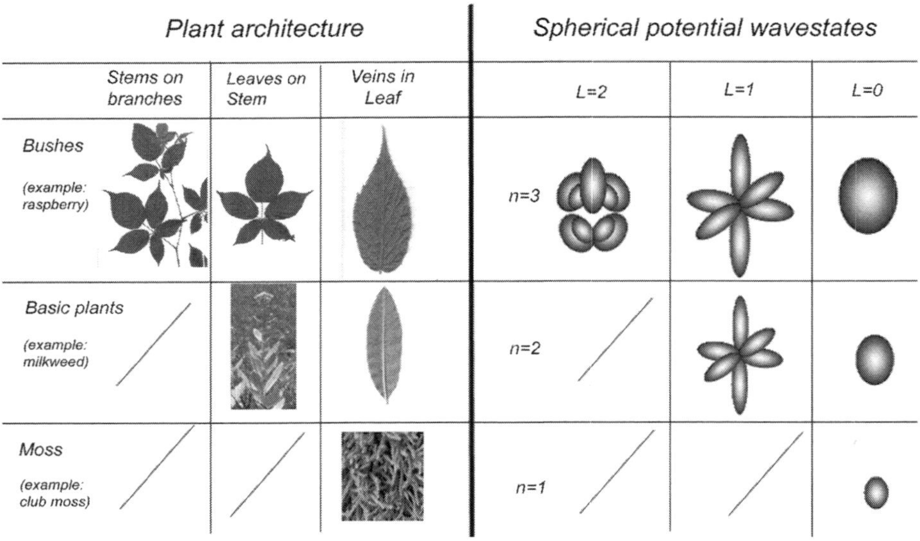

| Plant architecture | | | | Spherical potential wavestates | | | |
|---|---|---|---|---|---|---|---|
| | Stems on branches | Leaves on Stem | Veins in Leaf | | L=2 | L=1 | L=0 |
| Bushes (example: raspberry) | | | | n=3 | | | |
| Basic plants (example: milkweed) | / | | | n=2 | / | | |
| Moss (example: club moss) | / | / | | n=1 | / | / | |

*Figure 5.28*

However, even more profound connections become apparent when we expand our vision to consider whole plant architectures. In this chapter we've just noted that any particular plant has a number of possible levels of structural hierarchy, with specific organizations possible at each of these levels. Looking at plants this way means that each plant may be associated with a hierarchy of nested morphogenetic fields. This is in striking parallel to the hierarchy of nested wavestates that occur for a quantum particle in a spherically shaped energy potential. The parallel between nested metaphoric fields associated with a plant's full organization and nested wavestates for a spherical potential is shown in Figure 5.28. Note also that the complexity number is analogous to the set of quantum numbers, as both are used to specify the set of nested wavestates. Also, recall that in quantum mechanics, these nested wavestates are associated with energy.

Given the above parallels, it's apparent that we have on our hands an appropriate conceptual metaphor which uses features of quantum mechanical systems to account for characteristics of living things. What is interesting is that like all functional conceptual metaphors, there is not only a good analogy between the two systems, but a rather complete and systematic set of parallels that spontaneously become clear when the two systems are considered together. As is the case for all conceptual metaphors, this suggests that indeed, we have our hands on something which is a good fit — something with the capacity to yield a functional understanding of the situation.

Recognized fields in the physical sciences are associated with energy in one form or another. Also, as we've repeatedly pointed out, quantum mechanics wavestates are also associated with the total energy of the quantum particle. If this analogy between the quantum wavestate and the morphogenetic field can be so neatly drawn, we might at this point stop to consider what *kind* of energy might be associated with these morphogenetic fields and the biological forms manifested under their influence. Could it be that morphogenetic fields are intertwined with the age old, elusive concept of life-as-energy?

Thus, on the basis of the rather striking parallels between quantum mechanics and morphogenetic fields, let's complete the analogy by merely supposing the existence of an *emergent energy*. Let's postulate that this emergent energy arises from, and is supported by, the collective, interacting activities of cells, genes, and molecules, but that it in turn becomes an organizing force for the collective with its own systematic manifestations and characteristics. Let's propose that this *emergent energy* is what we often refer to as *life energy*.

## *Emergent energy*

A key hypothesis of this work is that the morphogenetic field is associated with an emergent energy. The concept of life-as-energy has been central to many ancient systems of thought, and continues to be used in 'non-scientific' discussions today. This notion of life-as-energy has arisen from our experience, and perhaps an intuitive grasp of the concept would be held by many. This is true of the historical recognition, and conceptual contextualization, associated with many forms of energy. For instance, no doubt you have an intuitive feel for kinetic energy (the energy of motion). If you've ever been hit in the belly by a moving sports ball, or fallen down the stairs, you know from first hand experience what kinetic energy feels like when it's transferred into your physical form. Everywhere we see things moving, or are ourselves moving — leaves rustling in the wind, our hands rising up to bring a mug of tea to our lips, a flag beating and wrinkling in the wind — we acknowledge a form of energy to be intrinsically associated with that motion. Having given it some thought, we have managed to recognize and contextualize kinetic energy, and now know how to manipulate and use it for our purposes. We can now design and build motors that will rotate the wheels of cars, the propellers of power boats, and the turbines of jet engines. We understand how these things move, and why. These experiences, and these conceptualizations of our experiences, make kinetic energy real to us.

I'm sure that drawing from our experiences, many of us have a similar intuitive knowing of emergent energy. Yet, it seems that our rational minds don't have a conscious, conceptual grasp of what emergent energy is, and therefore, we don't account for its effects or presence because we do not believe in it, we haven't conceptualized it. There is 'no such thing' as emergent energy. It is therefore not something that is considered a practicality when acting in the context of our bodies, our lives, our societies, and our living environments and supporting ecosystems. However, what if emergent energy were, like kinetic energy, a 'real world thing', with 'real world effects'? Perhaps we are like Madame Curie, who in working with invisible radiation from radioactive elements like radium, did not realize their potent effects, simply because radiation had not even been conceived of yet, and therefore, did not yet exist to humanity. Perhaps in today's conceptually dominated society, where everything is made to fit within our tried and tested rational systems, emergent energy is a crucial missing element.

## *More than a morphic metaphor?*

While some digging around in scientific journals is required to find them, and they aren't representative of mainstream life science, a number of very interesting reports suggest that morphogenetic fields/wavestates may not be merely metaphors, but rather, are entities with a completely tangible, measurable character. In fact, morphogenetic fields may be stuffless waves — entities that go by the more formal name of *electromagnetic fields* (we've discussed these in Chapter 4, pages 114–116).

In 2007, an *electric field* as strong as a lightning bolt was discovered within the bulk space of living human cells! Found to vary between 500,000 and 3,500,000 V/m, an electric field with such an enormous magnitude in the main region of the cell had not previously been known, or even expected, to exist.[6]

Electric fields contain energy, and therefore, the existence of a strong field within the bulk of the cell represents the presence of a small amount of previously unrecognized stored energy. Considering the human body contains about 100 trillion cells, the small amount of energy stored in the cell due to this incredible electric field adds up to roughly 37 Joules, which is about the amount of energy required to heat 1 ml of water up 9°C.[7] It's not much, but this stored energy is extraneous to the heat and chemical bonds which make up the bulk of the classically recognized energy of a life form. This 37 Joules of energy stored in an electric field within the cells of an organism could very well be the tangible measure of what we literally mean by the energy of life.

The existence and role of self-produced electric fields within an organism is slowly reaching the mainstream.[8] Recall from Chapter 4 that an electric field is produced wherever there is an unbalanced accumulation of either positive or negative charges. The body fluids of living creatures contain various salts, such as ordinary table salt (sodium chloride), dissolved in them. These salts are composed of bound atoms with electrical charge. Table salt's sodium atom is positively charged, while the chloride atom is negatively charged. This means the sodium atom is missing an electron, while chloride has an extra one. When dissolved in water, the charged atoms of the salt can separate to float about somewhat independently. A life form may separate the positive and negative charges of these dissolved salts (perhaps by pumping them into separate chambers) in order to create electric fields. While for some reason modern science has almost completely ignored their research, a number of scientists have published verified scientific reports describing the measurement of strong electric fields in trees, plants, fruits, salamanders, chick embryos, mice, rabbits, monkeys, and humans.[9] Even more interestingly, these electric fields change with the life form's

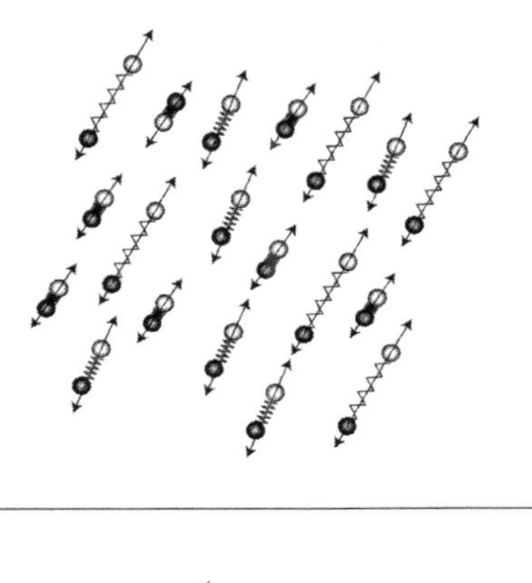

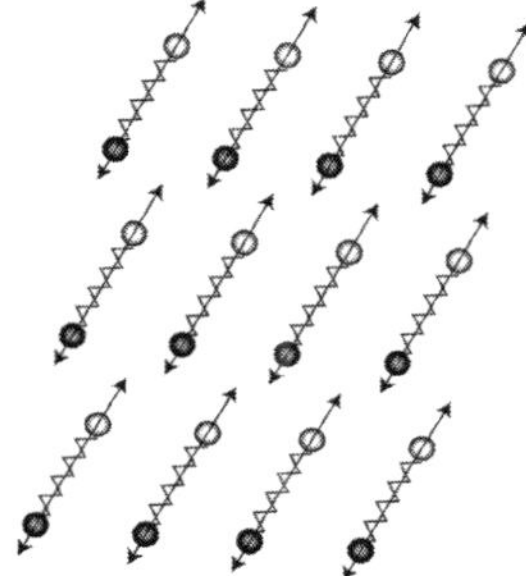

*Figure 5.29*

growth, with disease states like cancer, with normal events in the body such as ovulation, and with daily and seasonal rhythms. The specific way that these fields are generated and function in the life form remains unclear.

Importantly, the electric field of a life form has been shown to act just like a morphogenetic field! Alexander Gurwitsch, who believed the morphogenetic field to be an actual physical field (as discussed in Chapter 3, page 103), may have been right all along. In the regenerating planaria we mentioned earlier (see Chapter 3, and Figure 3.19, pages 100–101) an actual electric field with positive and negative poles oriented between the head and tail has recently been detected. While the mechanisms are unknown, it seems as though this electric field is what informs the planaria to regrow a head or tail at the right spot. If the head of the planaria is cut off and the stump exposed to an external electric field of opposite character, this can override the planaria's internal field to induce the incorrect formation of a tail where the head should be and vice versa. It's also been shown that in organisms that can re-grow a severed limb, an electrical field is present and implicated in the limb regeneration process.

Since 1968, with discussions becoming more serious in the past few years (2005 onwards), a few quantum physicists have been going on about electrically charged biological bits being able to enter a state of synchronized vibration which establishes a complete *electromagnetic field-wave* within a life form.[10] Many parts of any life form (such as water molecules, various proteins, cell skeletons, cell membranes) are electrically charged and are continuously vibrating with heat energy. Some quantum physicists have theorized that a long-range electromagnetic field can be created because the charged biological bits can establish a state of *coherent vibration*, where they are all moving the same way at the same time (Figure 5.29). Recall that electromagnetic fields are created whenever an electrically charged bit of matter accelerates. Well, vibration is one form of acceleration as the speed of the particle is constantly changing from high to low over the oscillation cycle. Associated with the vibration of just one charged bit of matter is an electromagnetic field-wave, which has a wavepattern directly related to the motion of the particle (Figure 5.30). The synchronous vibration of these electrically charged bits can create a long range field because of the way that the individual field-waves add up. If every vibrating bit is doing its own separate thing, even though the waves may be at the

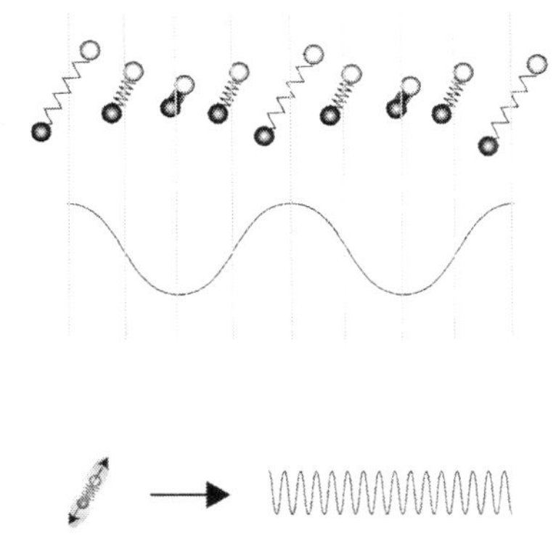

*Figure 5.30*

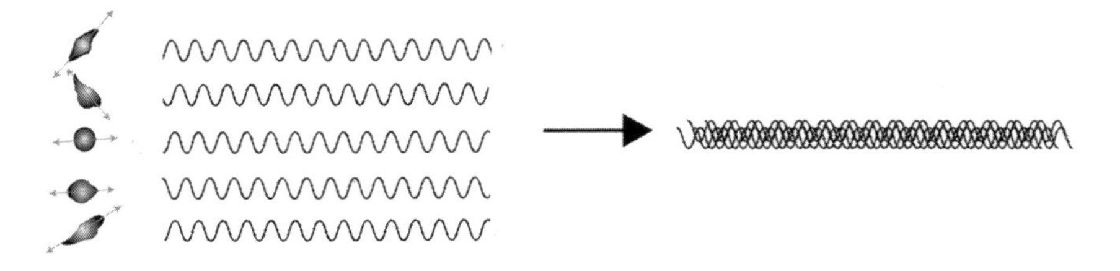

*Figure 5.31*

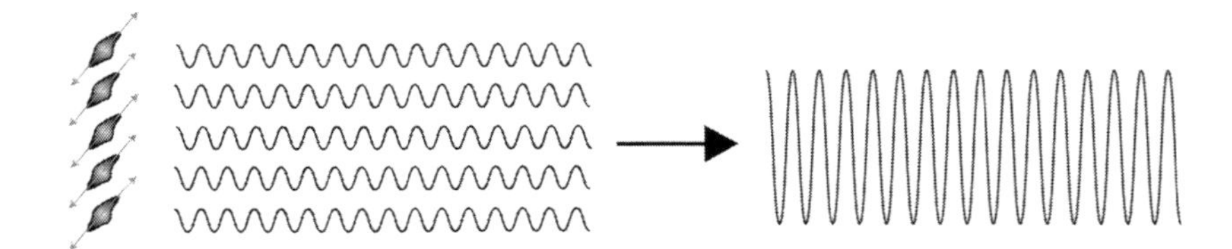

*Figure 5.32*

same amplitude and frequency, their patterns won't be matched up in time, and the resulting electromagnetic field-wave at the macroscopic scale will be a mix of little waves that are out of phase with one another (Figure 5.31). However, if all the vibrating bits are moving completely in phase with one another, they add up to form a singular field-wave which exists over the range that the charged bits are able to coherently vibrate (Figure 5.32). The coherent vibration of charged bits, and their associated electromagnetic field-wave, could span the dimension of individual cells, organs, and whole life forms.[11] Some of these physicists have been pushing the idea that this mechanism could generate an electromagnetic field within the life form which is responsible for its intricate large-scale organisation, and is even what we can call the morphogenetic field.[12]

There is already evidence that these large patches of coherently vibrating, charged biological bits exist along with their associated electromagnetic field-wave. Keep in mind that these biological bits are actually vibrating very fast, on the range of 1 million to 800 trillion times per second! The corresponding electromagnetic field-waves they produce would therefore be classified as radio-waves, microwaves, or light waves. Remarkably, radio waves have been detected from growing yeast cultures, microwaves have been reported from contracting frog legs, and there are many reports of light emission from a wide array of living things.[13]

That living things emit light, and appear to have great uses for it, remains the most well studied biological electromagnetic phenomenon as low levels of light are much easier to detect than high frequency microwave radiation. In fact, since the 1920s, light emission from living cells, tissues, and whole organisms has been quite extensively measured and characterized. The granddaddy of the morphogenetic field concept, Alexander Gurwitsch, was the first to postulate that living things use a special form of self-produced light in their self-assembly repertoire.[14] Working in the era before sensitive light detecting technologies had been invented (we have these now and they're called *photomultiplier tubes*), Gurwitsch ingeniously used living things as his light detectors. Gurwitsch noticed that in a sprouting plant, once the root begins to develop, the cells transition from a long quiescent period to a sudden-onset stage of rapid, coordinated cell proliferation. He proposed that self-produced light was the signal cells used to initiate the stage of co-ordinated cell division. Lacking a sufficiently sensitive light detector, he reasoned that if an actively proliferating onion root was producing light he could use it to illuminate a root that was still in the quiescent stage, and by doing so, tell the quiescent root to begin the stage of rapid proliferation. Separating the two roots by a glass chamber to ensure that no molecular messenger could pass between them, Gurwitsch proceeded to demonstrate that the illuminated place on the quiescent root began to rapidly proliferate, while non-illuminated portions, as well as a control root, remained in their original, non-proliferating state.

By the 1950s, Gurwitsch's work came to be scorned, as at the time there was no way to definitively attribute the observed occurrences to self-produced light from the organisms. This remained the case until the invention of sensitive light-detecting photomultiplier tubes. A number of recent investigations confirm that Gurwitsch was right — living things from bacteria to human beings are producing low levels of light.[15]

Living things appear to use self-produced light to coordinate growth events.[16] The light emitted from growing plant roots has now been detected by photomultiplier tubes. Yeasts, which are single-celled fungi that loosely associate with one another in a collective population, undergo a stage of coordination characterized by synchronous cell division. The yeast collective has been found to emit a pulse of light about an hour before this coordinated stage begins. Cancer cells are immortal cells which continuously divide and eventually take over and kill the rest of the organism they originated from. Remarkably, the light produced by cancer cells is orders of magnitude higher than the light produced from healthy cells. In a collective of healthy cells their emitted light saturates and begins to decrease as their population increases. In cancer cells, their emitted light increases dramatically with their increasing population. These examples of light emission from yeast, onion roots, and cancer versus normal cells indicate that cells may use light to coordinate a period of rapid proliferation. Exactly how this works is entirely unknown.

Yet there's even more to the story, as there's evidence that the light emitted from living things has the very special property of coherence, which supports the existence of a large zone of synchronously vibrating biological bits. The German scientist Fritz Albert Popp has spent thirty years measuring the light emitted from organisms.[17] He has shown that in plants, a whole leaf gives off coherent light. If the leaf is ground up, it still releases light, but the light is no longer coherent. Therefore, the properties of light emitted by leaves show evidence of a large scale electromagnetic field existing in whole organisms and depending on coherent vibration of biological bits. This means that light from living things appears to come from a coherent microscopic state of vibrating bits which depends on the integrity of the whole organism.

While morphogenetic fields/wavestates, emergent energy, and electromagnetic manifestations of life's energy and fields may have a number of important physical effects and properties, and have the potential to be explored in great detail, perhaps disappointingly, we will leave these for another time. We'll now step ahead to see what the implications the mere existence of emergent energy has on the properties and processes of collectives of living things.

# 6. Seeds for a New Ecology

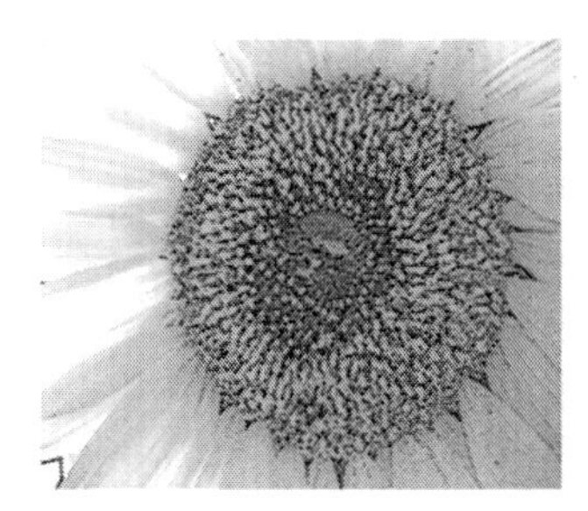

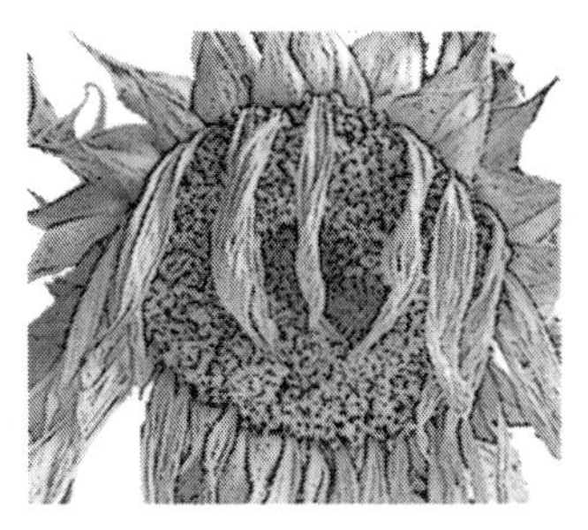

*Figure 6.1*

Everywhere we look, and in seeming conflict with the universal second law of thermodynamics, nature temporarily, but spontaneously and irrefutably, spurts, sprouts, unfolds, crawls, colonizes, wiggles and bursts towards various pattern states we intuitively recognize as embodying higher order. And yet, at some point in the life cycle of a living thing, the situation changes and the reverse is true. The form begins to slow, sag, ache, wrinkle, brown, and to eventually die and disintegrate entirely (Figure 6.1).

We shall see that if living things can be said to occupy discrete states of emergent energy, then this represents a new kind of system-of-systems for which conventional thermodynamics principles apply. Using surprisingly simple concepts, some spontaneous, time-dependent processes of living things can be effectively contextualized and understood. Ecology, among the most important and most neglected branches of science, currently resides in an inferior position to other 'hard' sciences as it lacks the underlying theoretical basis that other hard sciences (physics, chemistry) have.[1] The ideas outlined in emergent thermodynamics may be among the seeds of an entirely new, functional, theoretical basis for ecology.

## *The trouble with ecology*

Ecosystems are a hard thing for us to wrap our minds around. For one, they're a lot more expansive than the everyday human scale, and their progressions tend to take a lot longer than a human lifetime. Moreover, an ecosystem can't exactly be extracted for study under controlled, laboratory settings. Think of a forest, the green and leafy kind that grows in the North East United States. Of course in a forest there are many different kinds of trees — in this case we'll find maples, oaks, ash, basswood, pine, cedar, birch, and many others. Beneath the leafy canopy a vast array of smaller plants like wild raspberries, ferns, trilliums, and tree seedling are growing. There are mosses, lichens, and mushrooms growing on rocks, decaying logs, and on the tree trunks themselves. And there are many different kinds of animals in the forest. There are ants, centipedes, millipedes, leaf hoppers, cicadas, caterpillars, butterflies, bumblebees, praying mantises, fire flies, and mosquitoes. There are blue jays, cardinals, sparrows, and chickadees. There are snakes and frogs. There are chipmunks, mice, voles, deer, bears, and foxes. The soil is also alive with earthworms, fungus, bacteria, and yeasts. We mention but a few of the different entities living in this ordinary forest.

There are so many different kinds of individuals in a forest like this, but none of them exist in isolation. Rather, they're intricately interdependent upon one another via a web of complex relationships. The plants eat the energy of sunlight and remove carbon and nitrogen from the air in order to create their exquisite life forms. In turn, the plants emit oxygen for the other forest dwellers, and are a primary source of food and shelter for animal life. In turn again, the animals nourish plant growth with their carbon dioxide exhalations and enrich the soil for plants in death when their bodies decompose. The fox eats rabbits and is dependent upon them as a food source, but the rabbits benefit as a species, and the ecosystem benefits as a whole, by having rabbit populations controlled so they won't over multiply, eat too many of the plants of the forest, and ultimately face starvation. The relationships between the many creatures of the forest are extensive and intricate.

Ecosystems exhibit their very own characteristic progressions through space and time. These progressions are directly related to what's going on between all of the many individuals of the ecosystem, and obviously depend on the availability of essential energies, nutrients, and the absence of toxins that inhibit life. *Natural succession* is one example of such characteristic progression.[2] If an open plot of land exists with no vegetation on it (perhaps it was burned in a fire, or the vegetation was clear-cut and paved over by humans) the land will first grow mosses and lichen, and then grass and other simple plants. In time, the grassland will give way to a different pattern of vegetation characterized by simple plants and bushes. With more time, a first forest may develop, but in the end, this first forest may give way to a different type of forest characterized by entirely different trees and life forms. This final forest remains stable in its composition over a long span of time.

Eutrophication of an aquatic ecosystem is another kind of time-dependent progression the ecosystem may exhibit. In eutrophication, excessive amounts of plant-favouring nutrients are introduced to a waterway such as a lake. One may think that it would enrich the life of the ecosystem to receive additional food for growth. Yet the result of eutrophication is a whole bunch of dead fish! This unexpected result reflects the delicate balance of an ecosystem with all of its highly interdependent parts. The additional nutrients lead to excessive growth of the most basic forms of aquatic plant — typically slimy masses of algae — which run through quick life-cycles to decompose at the bottom of the lake. This massive growth of plants leads to a massive increase in dead plants, which leads to a massive increase in the number of detritus-eating microorganisms. The detritus eating microorganisms use oxygen to breathe, and their dramatic increase in numbers means the water becomes depleted of oxygen. This ultimately leads to a massive die off of larger animals like fish who need oxygen to survive.

Compare these features of ecosystems to those of a solar system, which consists of a sun orbited by a few planets. Even more so than the ecosystem, the solar system is much larger than the everyday human scale and its evolutions take much longer than the average human lifespan. In spite of this, it's relatively easy for us to account for the progressions of the solar system. This is because the motion of the planets can be reliably accounted for by an underlying set of consistent physical rules which are stated in terms of identifiable characteristics like the planet's mass, its distance from other planets, and the predictable effects of the force of gravity.

There is no overriding set of rules that we have yet come up with to account for the progression of the intricate webs of life represented by an ecosystem. At the moment, the very best we can do is see an ecosystem in terms of its individual life forms and the relationships between them. The closest thing to a theoretical framework in ecology is called *ecological network analysis*. This is a viewpoint related to systems science, which considers the giant webs of relationships at a stable stage at a snap-shot in time. Ecological network analysis works by focussing on the numerous exchanges of an essential nutrient (such as carbon, nitrogen, or phosphorus) or recognized forms of energy (sun, food, and heat energy) between the many members of the ecological community. An example of an ecological network for an extremely simplified, very coarse 'ecosystem' is shown in Figure 6.2. This figure indicates energy transfers between plants, herbivores (animals that eat plants), omnivores (animals that eat plants and other animals), carnivores (animals that eat other animals), and detrivores (animals that eat the dead bodies and solid wastes of other plants and animals). The ecological network tracks the flow of energy (or a single essential nutrient such as carbon) through the ecosystem as it passes from one organism

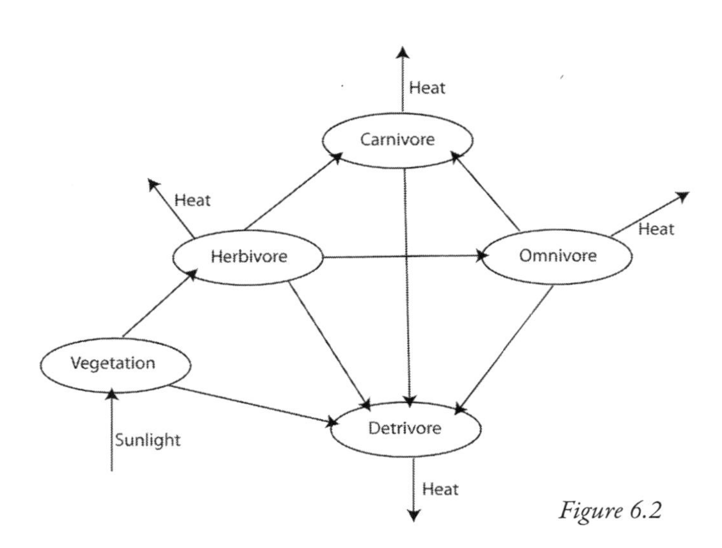


*Figure 6.2*

to another. Ecological network analysis allows us to estimate important things such as the stability of the ecosystem, its level of intricacy, its maturity, and its overall degree of health. It is also useful to identify cohesive cycles of relationship in the great ecosystem web.[3]

The problem with ecological network analysis is that it considers the ecosystem at only one particular snapshot in time, and is not able to map out how the system is to change in time. This is a problem as some of the most important things we need to know about ecosystems are how they will change in time, for instance, how an ecosystem will respond as a whole to the introduction of a new human-made pesticide. Because the ecological network is not designed to respond and change in time, it has somewhat limited modelling powers. For instance, ecological network analysis cannot account for the phenomenon of natural succession, or predict that a lake receiving fertilizer run-off may ultimately experience a large die-off of fish.

To model the time-dependent changes of an ecosystem requires a vast increase in the complexity of the model. However, the model typically becomes so complex that we don't get accurate solutions due to the assumptions and simplifications we need to make to get any solutions at all. A focus on a single factor at a single point in time is not enough to account for ecosystem progressions. A phenomenon such as eutrophication occurs because of time-dependent changes in the availability of essential nutrients like oxygen, phosphorus, and nitrogen. The model must consider rates of plant growth in response to all environmental conditions including the concentrations of phosphorus and nitrogen, rates of plant death, detrivore populations and their growth and death rates, detrivore oxygen consumptions, populations of all crayfish, clams, and fish and their oxygen consumption rates and base requirements. Taking a look at the basic factors that need to be considered for even the very simple five-part ecosystem, we see that to account for these additional factors, the model quickly mushrooms into a web of complexity as indicated by Figure 6.3. For an effective model, the mathematical specifics of each of these transfers must be accounted for. This becomes impossible for an ecosystem consisting of many hundred, or even many thousand relevant parts.

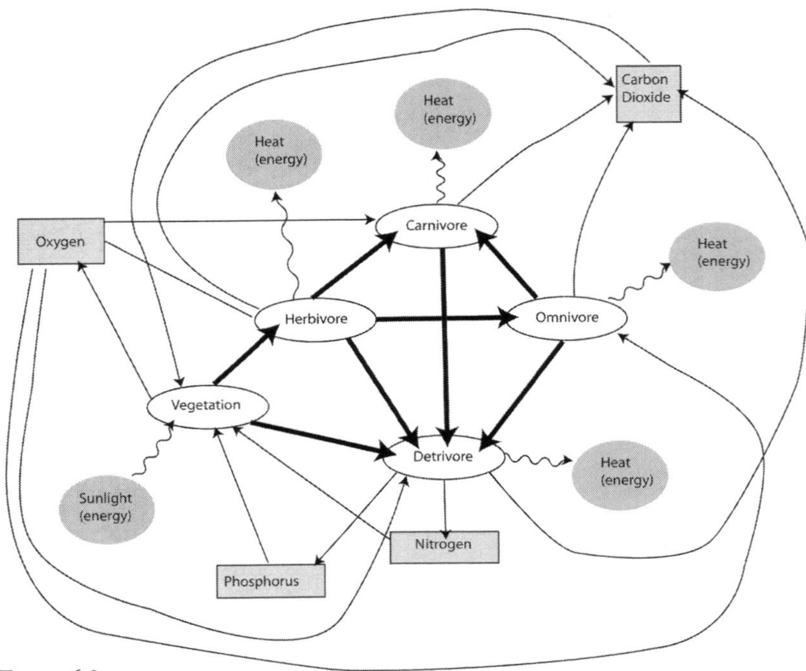


*Figure 6.3*

A cup of water is another system composed of a vast multitude of parts (water molecules) which goes through characteristic progressions in response to various changes in its environment. With increases or decreases in temperature, the water boils or turns to ice. With stomping footsteps, the water's surface vibrates as a wave. To account for what happens in this cup of water under these various circumstances, scientists don't keep track of what each and every molecule in the septillion molecules in the cup is doing. Instead, they define new parameters such as temperature, and new theories such as thermodynamics, or the dynamics of stuff-based waves, which are applicable to the cup of water as a collective whole.

We are going to follow suit to develop an analogy of thermodynamics applicable to an ecosystem. In this, individual organisms in their emergent energy states are going to be seen as analogous to molecules in their energy states. Surprisingly, we'll see that this analogy is a good fit and allows us to account for the spontaneous progression of an ecosystem (natural succession) and time-dependent perturbations such as the eutrophication of a lake. We then expand on the idea to get a grasp of why individual organisms necessarily progress through a characteristic life cycle.

## *Emergent thermodynamics*

In Chapter 5 we developed the concept of morphogenetic wavestates for plants. An implication of morphogenetic wavestates is that any plant existing at any particular region on Earth can be seen to represent an occupied state of emergent energy in a great plant kingdom wavefunction consisting of many possible, but not necessarily occupied, morphogenetic wavestates (this is shown in Figure 6.4). But in order to go any further with this concept, we must first be able to at least roughly estimate the amount of emergent energy associated with a particular biological form. In quantum mechanics the amount of energy associated with a wavestate increases as the structural complexity of the wavestates increase. In parallel to quantum mechanics, let's therefore propose that the amount of emergent energy of a biological form is directly related to the structural complexity of the system of morphogenetic fields associated with that biological form. We reasoned in Chapter 5 that the structural complexity of the morphogenetic field ensemble has two components: one corresponding to the number of levels of structural hierarchy, and one corresponding to the intricacy of patterning of a particular level. Thus, we would say that a maple tree is more complex than a grapevine, while both share the same leaf structure, and that a maple leaf is more complex than a blade of grass, while both are leaves. As it takes both the number of embedded structural levels and the complexity of each level into account, let's use the previously defined *complexity number* as a quantity that measures the plant complexity. Therefore, we are proposing that the amount of emergent energy associated with a particular biological form is related to the complexity number. We are using the complexity number in direct parallel to the use of a quantum number.

Next we must determine what is meant by the *degeneracy* of the emergent energy state, that is, the number of replicate levels existing at the same energy. In nature, there are great variations of pattern about an archetypal theme. In Chapter 5 we saw that while there were distinct pattern archetypes (that is, needle, blade, fishbone and rotator), there were many different variations in the morphogenetic field and the corresponding plant patterns for each archetypal pattern. For a particular morphogenetic field type, we will assume that these variations upon the archetypal pattern of a particular morphogenetic field type correspond to insignificant changes to the emergent energy. This can be seen as being similar to the case of a quantum particle in a spherically shaped potential energy, where the primary quantum number specifies the energy, and all nested wavestates, which are an assortment of different patterns, share the same energy (recall this material was discussed in Chapter 4, pages 121–124). For the case of biological form, the degeneracy of levels increases with the complexity of a morphogenetic field archetype. This is because many more variations of pattern are possible as the complexity of the archetypal field type increases. For instance the rotator field, which can vary the form of the central pattern, the angle of rotation, and the scaling of the central pattern, has a great many more variations than the needle field, which can vary only the length-width ratio. Therefore, the number of degenerate emergent energy states increases dramatically with increasing amounts of emergent energy (complexity). Again, these ideas parallel quantum mechanics for quantum particles in spherical potentials.

We can now conceive of a plant kingdom energy state diagram (as implied by Figure 6.4). Each empty emergent energy state (represented by an open __ ) represents a possible form that a plant species may assume. When a plant grows as a particular form, it has associated with that form a morphogenetic field with a corresponding amount of emergent energy. The amount of emergent energy is given by the complexity number. The actual plant represents the occupation of that particular emergent energy state in the plant kingdom energy state diagram.

In our walk through the mind garden (Chapter 4, pages 127–137), we have seen that for collective systems of individual things such as

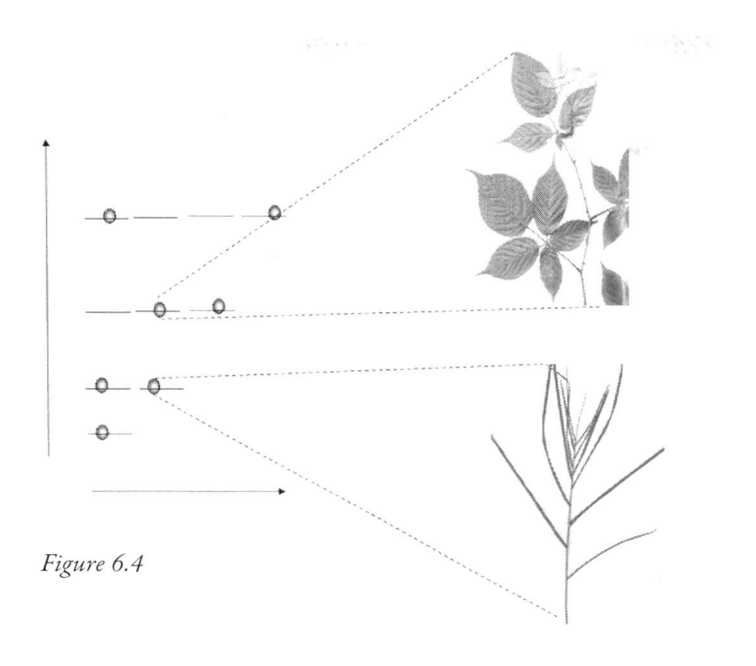

*Figure 6.4*

electrons, atoms or molecules, which each occupy energy states, there are certain laws of the universe which indicate how the particles will be probabilistically distributed amongst these energy states. We have seen that new parameters can be defined that account for the behaviour of the collective system. The most basic of these new parameters are heat, temperature, and entropy. Recall that there is a statistical definition for temperature. At low temperatures, the particles are found sharply distributed in the low energy states. At very high temperatures, rather than being distributed sharply in high energy states, the particles are spread throughout different states, and there is an equal probability of finding them in any state, high or low. These aspects of temperature are shown in Figure 4.29 and Figure 4.30, back in Chapter 4, page 136.

Now, in any place where a community of plants exists, the above ideas can be used to estimate the amount of physical plant form (biomass) growing at particular levels of emergent energy, and to develop an analogy to thermodynamics that is exclusive for systems of living things. Here's the kingpin idea: if living things occupy states of emergent energy, then the universal laws of thermodynamics should apply to how these emergent energy states are occupied in time, just as they do in non-living systems. If we understand plant forms to occupy emergent energy states as such, then an *emergent temperature* (emergature) can be defined for the region where a community of plants is growing. This emergature is based on the distribution of the plant mass amongst the possible emergent energy levels (morphogenetic wavestates). Thermodynamic temperature is specific to the distribution of mass across various *kinetic energy states* for a collective of individuated atoms or molecules, while emergature is specific to distributions of a collective of individual living things within *emergent energy states.* We've already mapped out a system of morphogenetic wavestates that can be used for different plant types and a corresponding complexity number to roughly quantify the amount of emergent energy of a particular plant (this happened in Chapter 5, pages 173–175). Now, given a patch of land with plants growing on it we can construct a rough energy density diagram that shows the distribution of emergent energies for the plants growing in that particular region. From the energy density diagram we create we can see if the idea of temperature, as represented in Figure 4.30, complies as an emergature specific to the distribution of plants in their emergent energy levels for any particular patch of land as a function of time.

Let's see how these concepts apply to an actual situation. In order to do this, we would ideally like to find a patch of land that's been cleared of all of its vegetation, and wait around taking detailed observations of the plants that grow on it for about one hundred years or so. This would give us an idea of how the emergent energy density of the area changes as a function of time. As I don't have this kind of time, we're instead going to consider three plots of land that have been culled of vegetation at subsequently more distant points in time. These areas can be seen as snapshots of what we would most likely see if we were waiting around and watching for a long time. So let's consider the plant distributions in a recently disturbed region we'll call the *grassland* which was disturbed by mowing approximately two months prior to investigating it; a nearby region we'll call the *meadow,* which has remained undisturbed for approximately thirty years; and another nearby region we'll call the *deciduous woodland,* which has remained undisturbed for approximately seventy-five years.

Let's begin by examining the recently cut grassland (Figure 6.5). This grassland was found populated mostly by grasses (complexity number 32) and, to a lesser extent, by simple plants such as ragweed (complexity number 33), milkweed (complexity number 33), vetch (complexity number 33) and clover (complexity number 33). This grassland was once a deciduous forest, which characterizes the equilibrium plant community for the region, but had been cleared many decades prior to this inquisition into its present plant life.

The graph on the right of Figure 6.5 is the energy density diagram for the plants in the grassland. The energy density diagram plots the estimated amount of plant growing at a particular plant complexity number (emergent energy). The information for the graph shown here, and ones shown in the following pages, was obtained by masking out areas of photographs (similar to the one shown to the left) for plants of a particular complexity number, and summing up the masked area of plant growth at each complexity number with a computer. The results are averages of four pictures of different areas of the particular growth regions (grassland, meadow and deciduous forest). The axis of the graph shows the range of all possible complexity numbers for known plants, under the current (very rough) system of morphogenetic wavestates that have been defined in this work.

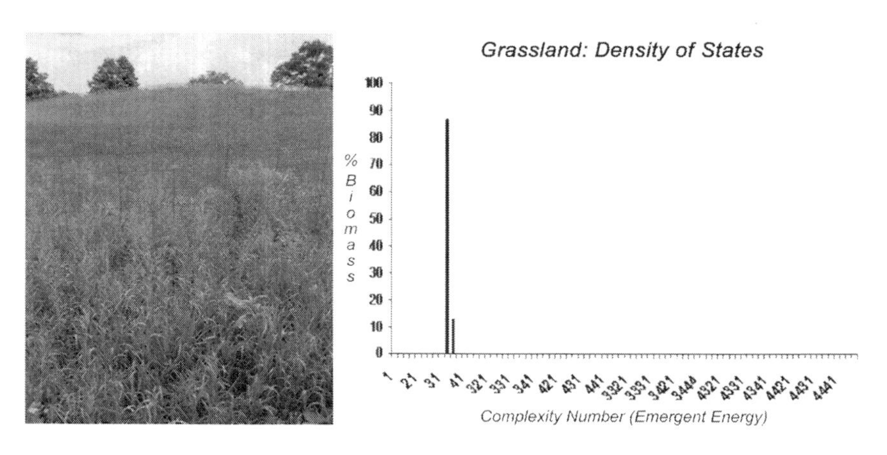


*Figure 6.5*

We can do the same thing for the meadow, a region near the grassland, which had been deforested some years past, but allowed to re-grow for approximately thirty years (Figure 6.6). This meadow was found to be populated by sumac (complexity number 333), wild plum (complexity number 3333), wild grape (complexity number 34), milkweed (complexity number 33), ragweed (complexity number 33) and grasses (complexity number 32).

Finally, we can examine the plant growth in the deciduous woodland, which represents a plant community that is close to the equilibrium community for the area (Figure 6.7). The deciduous woodland was found populated by mosses (complexity number 2), ferns (complexity number 33), trillium (complexity number 32), purple bush raspberry (complexity number 334), ash (complexity number 4333), beech (complexity number 3333), birch (complexity number 3333), oak (complexity number 4333) and maple (complexity number 4334). This deciduous forest was located within 2 km of the grassland and the meadow, and therefore represents a similar growing region. This forest has remained undisturbed for approximately seventy-five years.

In the plant communities we have just studied, and in parallel with the basic concepts of thermodynamics we have covered, we can understand that the entropy, and the corresponding emergature of the patch of land, is spontaneously increasing as time passes. After clearing the

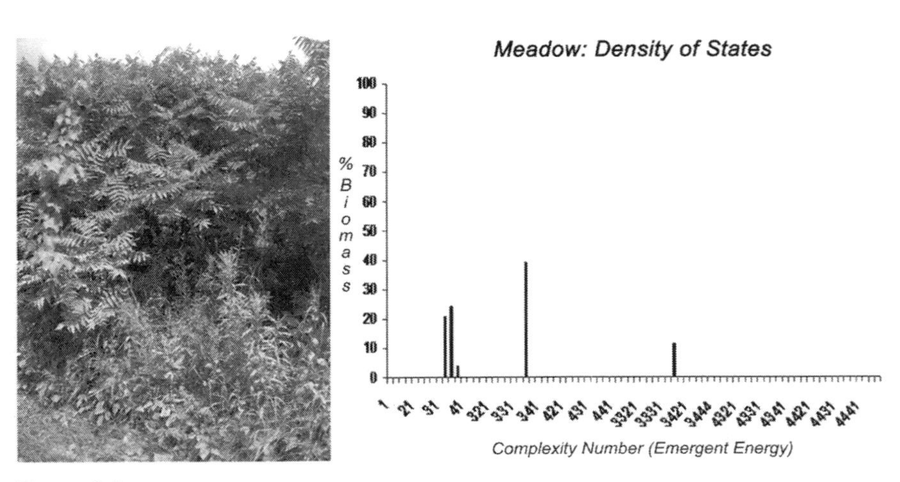


*Figure 6.6*

land at some point in the past (which is effectively a zeroing of the emergature of the area as most plants are removed), the immediate regeneration is as grassland, which represents a large amount of biomass with relatively low complexity. The distribution of biomass in the grassland is predominantly, and sharply, distributed in lower emergent energy states (Figure 6.5). If we compare Figure 6.5 with the top image of Figure 4.30 we can see a distinct similarity. Thus, in direct analogy to thermodynamics, this similarity implies that the emergature of this grassland is low. In contrast, the mature deciduous woodland contains a much wider range of complexity, with a distribution from moss to maple trees, and a peak in biomass at a much higher level of emergent energy than found in the meadow or grassland (Figure 6.7). Figure 6.7 can be compared to the bottom image of Figure 4.30. This represents a system of much higher emergature and entropy to that of the grassland. Consistently, the emergent energy distribution of plants of the intermediate, limited re-growth meadow, are indicative of an emergature somewhere in between that of the grassland and mature woodland (Figure 6.6).

After clearing the land, the plant community has spontaneously progressed from grassland, to meadow, to deciduous forest. This is a spontaneous progression from a state of low to high emergature. In ecology, this spontaneous evolution of a cleared region of land through

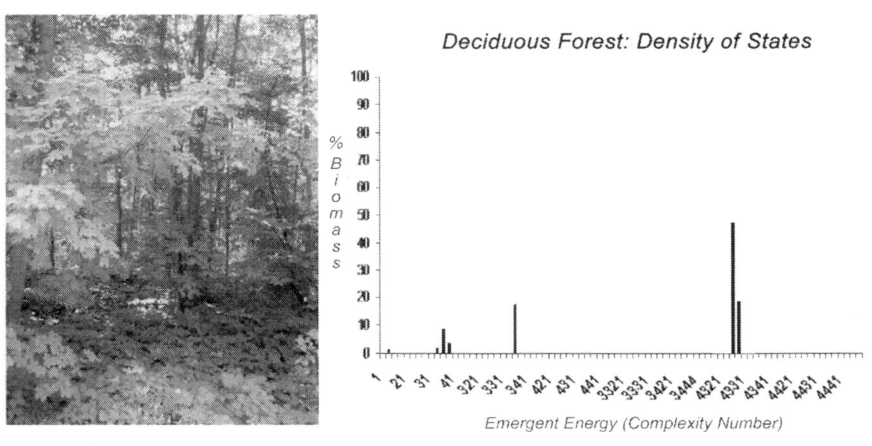


*Figure 6.7*

characteristic stages on its way to an equilibrium community is a well known phenomenon known as *natural succession*. That the progressions and processes of living systems can be seen to follow the same laws as non-living systems in their recognized emergent energy states is greatly intriguing as it implies that the concepts of morphogenetic wavestates and emergent energy behave self-consistently, and in accordance with previously derived scientific laws. It opens up the possibility of understanding, predicting and possibly mediating many biological processes, such as natural succession and the life cycle of an organism (we'll get to this soon), from a very straightforward, manageable, and holistic perspective. Moreover, as this basic theoretical perspective of living things existing in states of emergent energy was able to account for the well recognized real-world process of natural succession, this effectively supports the existence and real-world relevance of emergent energy.

## *'Thermal equilibrium' of plant communities*

With the application of a heat source, the temperature of a closed, insulated, non-living system-of-systems will increase with time as the object heats up, generally up to an equilibrium temperature which is a balance of heat input, to heat dissipation (Figure 6.8).

An analogous situation is proposed for a living system-of-systems, such as the plant communities we have been discussing. However, for the case of plant communities, 'heat' sources are proposed to be sources of matter or conventional energy relevant to life. For plants these are things such as: sunlight, moisture, physical temperature, nutrients, and carbon dioxide gas. In direct analogy to inorganic systems, supplying an 'insulated' ecological system-of-systems with emergent heat will lead, over time, to an increase in its emergature. Considering the spontaneous progressions of plant communities such as the ones we have just examined, note that this is exactly what is observed in natural succession. With continuous input of emergent heat sources, the low emergature grassland spontaneously progresses to a high emergature forest.

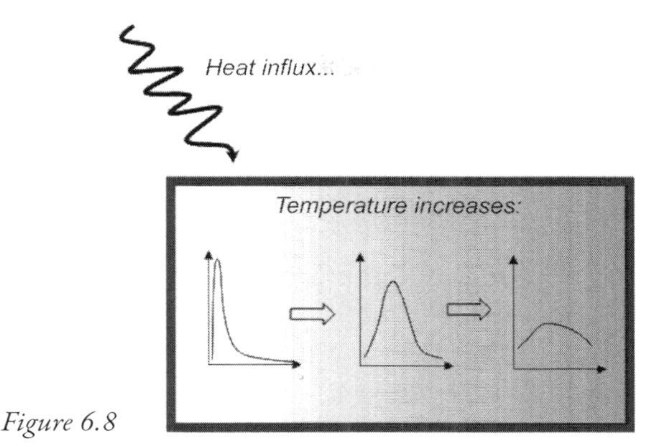


*Figure 6.8*

In analogy to a heated non-living counterpart, a collective of living things should also reach thermal equilibrium. Different regions of the Earth supply varying levels of emergent heat, as represented by the levels of sunlight, free moisture, soil quality and carbon dioxide. This means there should be different equilibrium states for plant ecosystems, depending on where they exist on the Earth. This is of course seen in reality. The seasonal levels of sunlight, the atmospheric temperature, and the available water present in the far north of the Canadian Shield are quite low in comparison to those of the Amazon rainforest. Correspondingly, the emergent heat sources of the Canadian shield result in a community of plants with a low equilibrium emergature, such as a boreal forest, where the most complex species, the pine, has lower complexity/emergent energy than other possible trees. In contrast, the higher levels of emergent heat in the South American Amazon rainforest create a community of plants with a much higher equilibrium emergature than the boreal forest.

## *Insulation*

Some systems-of-systems, such as an old growth forest, exist as complex self-sustaining communities. They have developed over long time periods, typically from a non-vegetated landscape such as a bare rock base, and have progressed from simple to complex by passing through phases such as lichen covered rock, grassland, bush, and particular forests. The characteristics of these self-sustaining systems-of-systems are that as time progresses, more nutrients and support for future generations are established. Dead organic plant matter, which represents sources of carbon and nitrogen that have been fixed from the atmosphere, provides additional nutrients for future plant generations and provides a basis for which plants can root and stabilize themselves. Intermediate stage plants such as bushes condition soil and supply shelter for their successors. The community is able to provide for its own needs, and has an abundance of relevant resources which provide additional support for generations to come. Over time the community grows in complexity to a stable equilibrium state which is ultimately conditional on the supporting characteristics of the environment.

In contrast, another type of system-of-systems is represented by mould growth on a peach. This community of mould is parasitic, and ultimately not self-sustaining, as its expansion and growth depend on the consumption of a limited resource (the peach). In a parasitic system, as time progresses, less and less nutrients and support for future generations remain. Interestingly enough, in contrast to natural succession occurring in plant communities discussed above, while the mould community increases in numbers, the increase to states of higher complexity (emergent energy) is not observed.

The thermodynamics concept of insulation assists in contextualizing these different phenomena in living systems-of-systems. Let's define a completely insulated emergent system (sustainable) to be one that produces new resources for future generations in the process of its growth. In contrast, a completely non-insulated system (parasitic) depletes nutrients from its non-renewable, supporting sources. We can understand that it is the differences in the insulation of these two living systems-of-systems which dictates the response to the sum of incoming heat sources. If it is an 'insulated' (that is self-sustaining) system-of-systems and heat is applied, the emergature of the system will increase in direct proportion to the amount of emergent heat applied. If however the system is non-insulated (parasitic), then the emergature of the system-of-systems will only increase to the extent of which emergent heat input is greater than the emergent heat dissipated. These ideas are shown schematically in Figure 6.9.

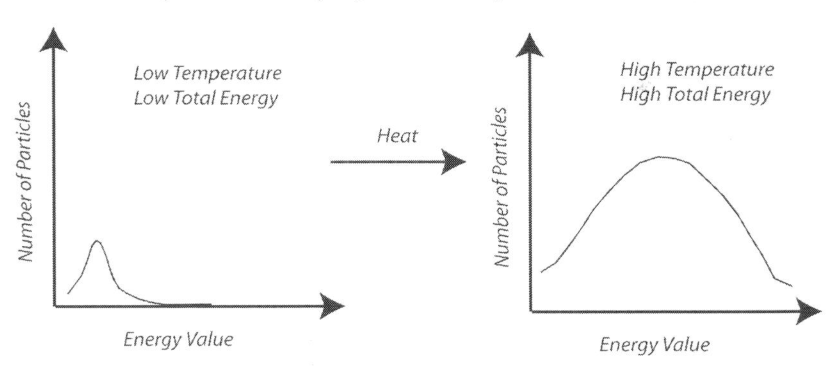


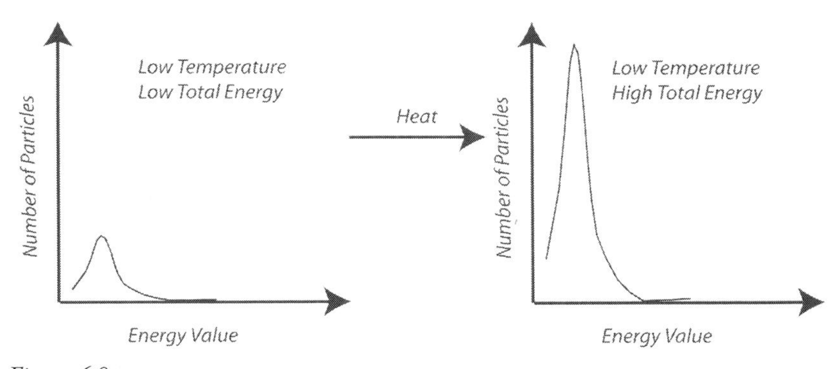


*Figure 6.9*

## *Emergent systems*

Let's attempt to further clarify the basic concepts of emergent thermodynamics by considering conventional closed and open systems (first discussed in Chapter 4, pages 137–139). In thermodynamics, a closed system is designated as one that is able to exchange energy, but not mass, with its environment. In contrast, an open system is one that continuously exchanges energy and mass with its environment. Closed systems ultimately reach thermodynamic equilibrium, representing a state where all macroscopic parameters (such as temperature, positional segregation) have become homogenized throughout the system, so that entropy is maximized. Thermodynamic equilibrium can be thought of as the complete fading out of any inherent patterning in the system. Therefore, a closed system cannot store a pattern for any prolonged length of time.

Open systems (Figure 6.10), due to the continuous exchange of matter and energy, do not reach thermodynamic equilibrium, but rather, operate in steady states of dynamic, pattern-based equilibrium characterized by constant through-flows of matter and energy. A living thing is a prime example of a dissipative open system. A living thing takes in sources of matter such as gases, water, and food-stuffs and exchanges different gases and biological wastes to its environment. You, a mammalian form, ingest streams of matter such as oxygen, water and sugar, and emit streams of matter such as carbon dioxide and biological wastes. You emit energy in the forms of kinetic and heat. You operate far from thermodynamic equilibrium with your surroundings, with a body temperature that is maintained at a nearly constant 37°C, and maintain an intricate patterning. Due to their inability to reach thermodynamic equilibrium, and the inherent complexities associated with continuous flows of matter through a system, the basic, conventional laws of thermodynamics must be restated in order to be applied to open systems. We discussed these ideas in Chapter 4, pages 137–142.

An even more sophisticated system context can be defined for living systems. In emergent thermodynamics, the steady-states of dynamic equilibrium of the living, dissipative, open system-of-systems are also representative of discrete states of an emergent energy, which is one morphogenetic wavestate in a grand morphogenetic wavefunction. Therefore, superimposed upon the open system operating far from thermodynamic equilibrium is an *emergent system*, consisting of a set of emergent energy states that can be occupied (Figure 6.11). The living system can be simultaneously seen as a dissipative open system operating far from thermodynamic equilibrium, and as an emergent closed system which can reach a state analogous to thermodynamic equilibrium. Note that the thermodynamic equilibrium of the emergent closed system is with respect to the distribution of occupied emergent energy states.

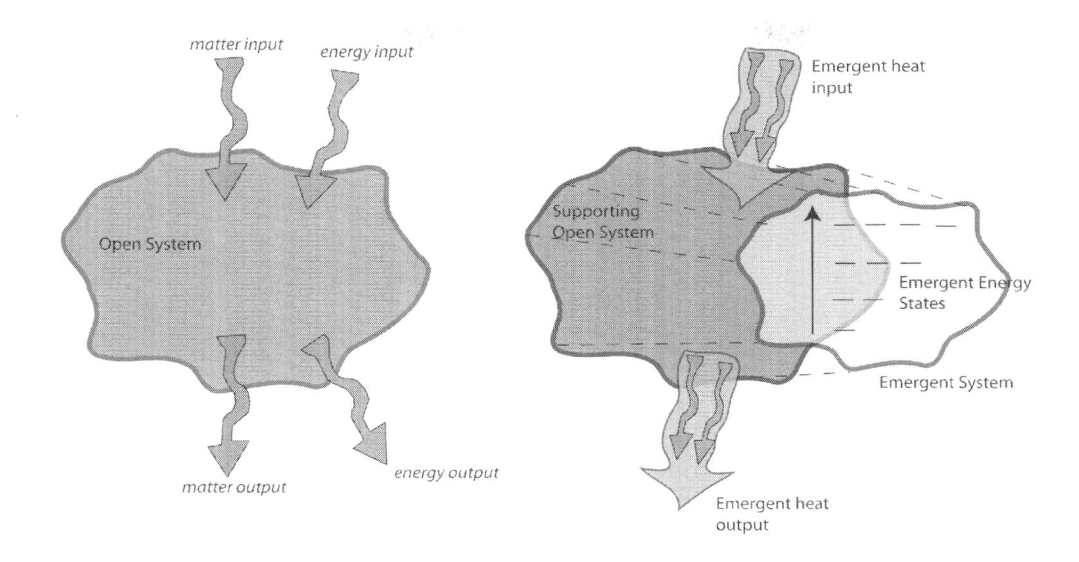


*Figure 6.10*

*Figure 6.11*

The living open system supports a closed system as, with the introduction of emergent energy states, material flow through the system can be re-expressed as 'heat' inputs to the emergent closed system. Recall that closed systems can accept energy exchanges such as heat, but not mass flows. This re-expression is possible as matter is involved in creating biological form, and these biological forms represent emergent energy. An emergent heat input is defined as any source of matter or energy that can increase the occupation of available emergent energy levels.

Energy and mass outputs of the emergent system fall into two categories: those relevant to the system as new input, and those that are true waste-products not recycled through the system. The emergent system can be further classified as *insulated* if de-occupation of an emergent energy state (death and organic waste) represents a recycled emergent heat input to the system (Figure 6.12). For instance, in most plant communities, dead organic matter represents sources of carbon and nitrogen that have been fixed from the atmosphere and additional nutrients and provisions for future plant generations. Over time the community grows in complexity to a stable equilibrium state that is ultimately conditional on the supporting characteristics of the environment.

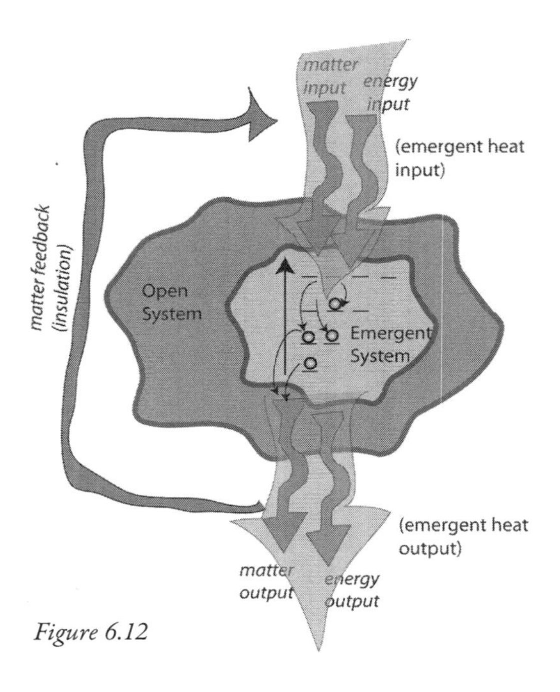


*Figure 6.12*

An emergent system with energy or mass outputs that are not relevant to further growth and development (that is, are not emergent heat inputs) can be classified as a *dissipative* or *parasitic* system, akin to mould growth on a peach. The emergature of an insulated system increases with emergent heat input, leading to a wider distribution of occupied emergent energy states. Conversely, the emergature of a dissipative system does not increase, leading only to increases in biomass in lowest emergent energy states, with eventual crash of the non-insulated emergent system when resources are all used up.

## *The conservation of emergent energy*

For this next part we're going to widen our definition of emergent energy states to include life forms other than plants. This will allow us to consider a more whole ecosystem.

Unfortunately, I haven't worked out a system of complexity numbers/ emergent energies for life forms such as insects, fish, frogs, and mammals.

However, I'd argue that based on the characteristics of their physical forms and other factors, these creatures are not necessarily of higher emergent energy than various plants we've discussed. Rather, I'd predict that bacteria would be the lowest emergent energy, while different kinds of yeasts, mushrooms, insects, and animals would be distributed within the various levels of complexity we've already mapped out for plants. The best intuitive guess for the emergent energy distributions of non-plant versus plant life forms is mapped out in Figure 6.13. Single-celled yeast would be analogous to algae; earthworms to moss; insects, crustaceans,

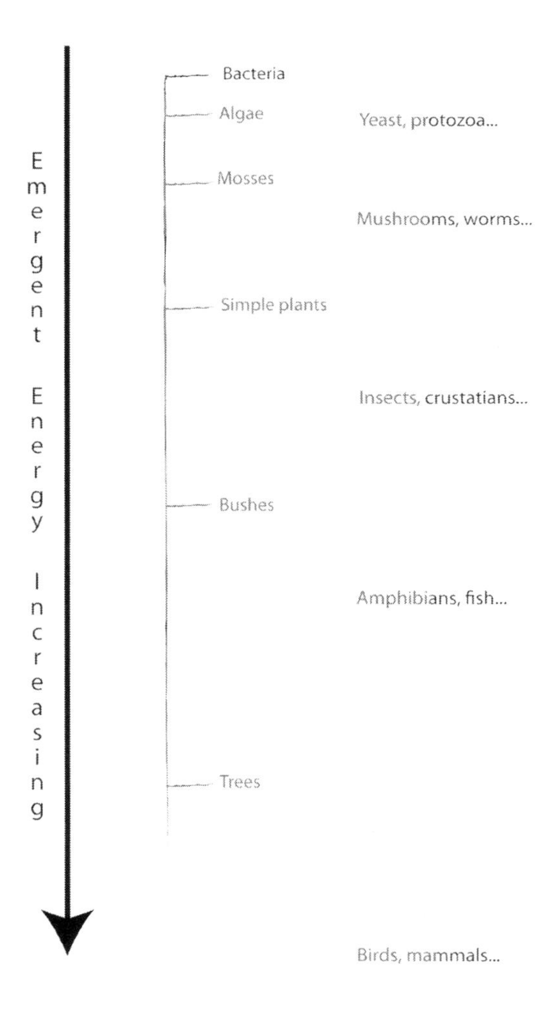


*Figure 6.13*

and mollusks distributed around basic plants; fish and amphibians around shrubs; birds and mammals at or higher than trees. The point is that in a state density diagram for an ecosystem, we'll consider populations of plants and animal species mixed amongst one another.

So far, we've been focussing on the *distribution* of occupied emergent energy states over a long time span on the order of one hundred years or more, where changes are occurring very slowly over the time span of a few decades. But there's something else to pay attention to, and that's the total amount of emergent energy as a function of time. As we saw previously, we can prepare a state density diagram for the plot of land. The distribution of occupied emergent energy states is indicative of the emergature (emergent temperature) which tells us how mature the plot is. However, there's another parameter we've not paid attention to, and that's the total amount of emergent energy for the whole plot of land. Given a state density diagram, the total amount of emergent energy can be estimated by multiplying the value of an emergent energy state by the amount of biomass in that energy state and adding up the calculations for all occupied energy states. If we start off with, say, a parking lot, we can clearly see that it has virtually zero emergent energy. If it's left unbothered for a few decades, the parking lot may well have transformed into a plot of knee-high grass and simple plants. The knee-high grass and simple plants of the parking lot represent a significant amount of emergent energy in comparison to the bare parking lot. Yet, in one hundred years, that undisturbed grassy plot will likely develop into a woodland, with 20 metre high trees, 20 cm in diameter, and an understorey of bushes, ferns, fungi. The forest represents a tremendous increase in the total amount of emergent energy in comparison to the grassland, as not only have higher emergent energy levels become occupied, but the amount of biomass in them is very high as well. This is schematically depicted in Figure 6.14. Note that in the examples of grassland (Figure 6.5), meadow (Figure 6.6), and forest (Figure 6.7) that we've just examined, we depicted the percentage of biomass in each emergent energy state, not the total amount, so each of these diagrams don't give any indication of the total emergent energy present.

In any case, it's intuitively obvious that a grassland has much, much less biological material than a mature forest, and that this biological

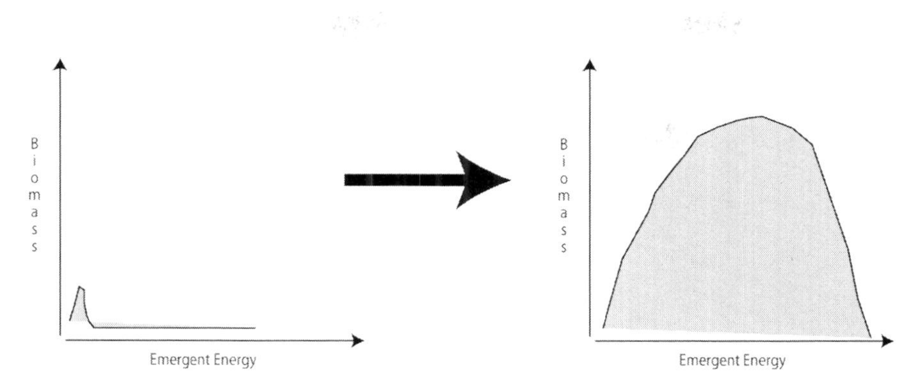


*Figure 6.14*

material is indicative of the total amount of emergent energy in the ecosystem. It's also obvious that this total emergent energy builds up very gradually with time, taking many decades to increase a noticeable amount. It's so slow that in any given year you'll probably not really notice any change happening at all. Since there's so little change in the total emergent energy of the system over a very short time period (a few months to a few years) we could consider the total amount to be constant on the short term. This is the first notion we'll need to keep in mind: the notion of conserved emergent energy in the short term.

Now, the basis of emergent thermodynamics thinking is that the emergent system responds to an emergent heat input by filling the lowest energy levels first. We don't see a forest of oak trees springing up in the parking lot. The changes occur through a series of stages reflecting a slow increase in emergature and total energy over a long time.

So what happens if we have a fairly mature ecosystem with, just to put a number on it, 100 units of total emergent energy (Figure 6.15-A). Now we go ahead and perturb this ecosystem by suddenly adding in a very large amount of emergent heat. This might be a big dump of synthetic fertilizer into a lake. Let's say that this spike of emergent heat is of sufficient strength that it would make a beginning stage system want to put a total of 50 more units of total emergent heat as occupied emergent energy levels (see Figure 6.15-B). If this heat source had been present and constant for the whole time that the mature ecosystem

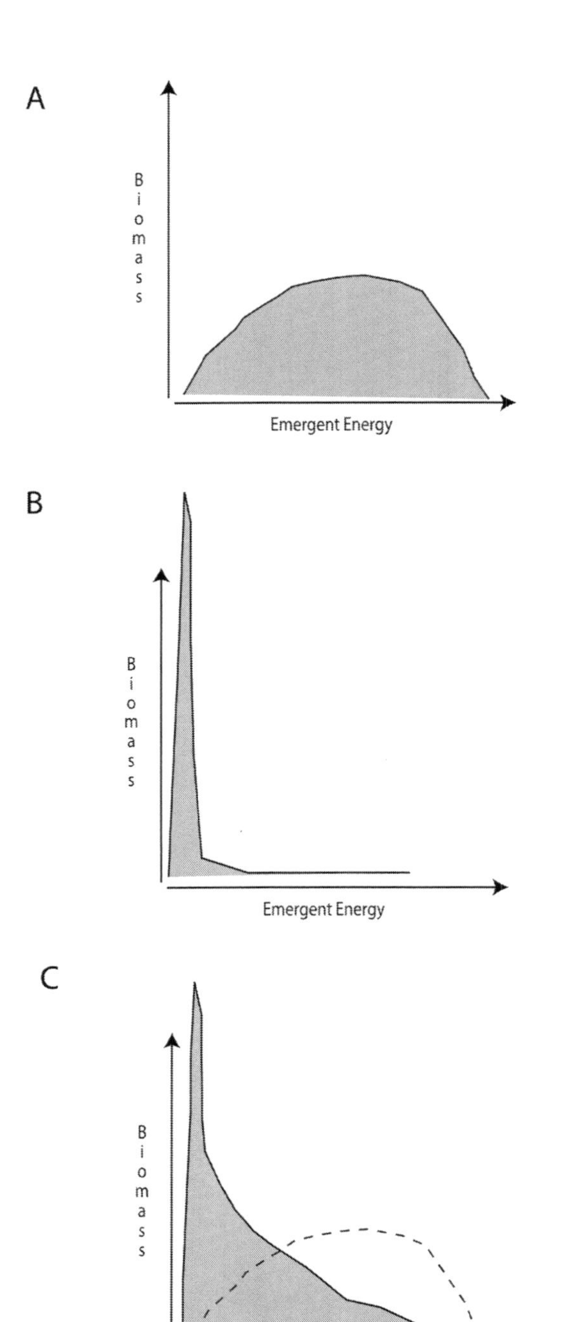
A
Biomass
Emergent Energy
B
Biomass
Emergent Energy
C
Biomass
Emergent Energy

*Figure 6.15*

developed, it would have lead to a mature ecosystem with a higher emergature and a much, much, much higher total emergent energy. Yet now this increased emergent heat is happening all of a sudden, and to the mature ecosystem. What do we expect to happen?

An interesting situation is established with this massively large emergent heat spike, for the system is driven to respond by creating an increased occupation of the lowest emergent energy levels, but the total amount of emergent energy cannot suddenly increase by 50 units in one year. So where will this additional 50 units of emergent energy come from? It'll come from a de-occupation of the existing emergent energy states. In this example, let's specifically consider a lake receiving a dump of synthetic fertilizer. In the lake, the life forms increase in emergent energy from the simplest to complex: from bacteria, to algae/ yeast, to basic plants/insects/crustaceans, to fish. Therefore, in order for the lowest emergent energy levels to fill as quickly and as high as possible under the influence of the strong emergent heat perturbation, de-occupation of the more complex levels such as fish, occurs. With more dead fish bodies around, more detrivore bacteria can grow, in addition to the growing algae, leading to a dramatic increase in the lowest emergent energy levels. This is shown schematically in Figure 6.15-C. In other words, the mature emergent system responds to very large and sudden additions of emergent heat as if the total emergent energy were a constant. This is of course what's observed in the case of a eutrophied lake.

From a systems viewpoint, this represents an unbalancing and/or a breakdown of the various checks and balances in place which keep the ecosystem in a dynamic state of equilibrium. While further work is required to substantiate and expand upon the dynamics described above, it seems as though an emergent thermodynamics viewpoint can assist in accounting for the overwhelmingly complex, time-dependent processes of an ecosystem.

## *Death. And the life cycle*

*The Life/Death/Life nature is a cycle of animation, development, decline, and death that is always followed by re-animation. This cycle affects all physical life and all facets of psychological life. Everything — the sun, novas, and the moon, as well as the affairs of humans and those of the tiniest creatures, cells and atoms alike — have this fluttering, then faltering, then fluttering again.*

**~ Clarissa Pinkola Estes**

Why do things spontaneously age and die?

Even nature appears to only temporarily 'cheat' entropy. Inevitably *all* things age, losing colour, shape and vitality, to eventually lose life itself, so that form breaks apart and dissolves into the originating landscape (just like the form formerly known as crow in Figure 6.16).

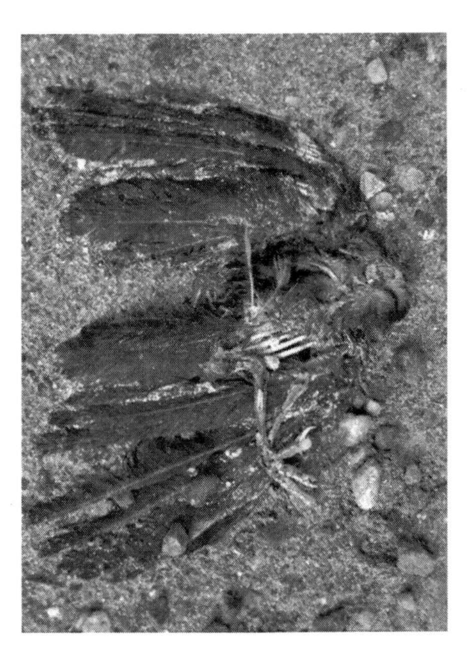

*Figure 6.16*

An outcome of emergent thermodynamics is the demonstration that all of nature's spontaneous processes, from the sprouting of the seed, the eager vibrant growth of the seedling, the development of the mature plant, the weakening of the form, the browning and loss of life from the form, and the denigration of the form into disassociated primary elements, are part of a natural life-cycle, in which *increases* in entropy occur all the way through. Again, we are speaking with respect to distributions of emergent energy states of the systems which constitute the system-of-systems.

We can get a better grasp of the life cycle as a phenomenon of entropy by considering a common illustration of the second law of thermodynamics — that of a drop of dye mixing in a bowl of water (which is a closed system after the dye enters the bowl). While it is certainly true that a drop of dye will not remain as a concentrated, localized mass in a bowl of water (a state of low entropy), but will instead spontaneously spread homogeneously throughout the water (state of maximized entropy), the progression is actually not as clean and straightforward as is often taught in school. In fact, the phenomena of dye spreading through water is, due to the many non-linear influences such as the interactions and relationships between dye and water molecules, convection currents of the fluid, and perturbing influences such as small vibrations of the table on which the bowl of water sits, a complex or chaotic system. In fact most dyes do not spread evenly into the water, but progress through states of complex chaotic pattern before the ultimately dispersed, homogeneous end state.

In this simple experiment, which is documented by the images of Figure 6.17, a drop of dye is added to a still bowl full of water. The drop of dye begins as a small, localized, low entropy configuration in the liquid, but it spreads rapidly (within 10 seconds) to form a complex streaming pattern with a form that stabilizes within about a minute. Subsequently, the pattern changes much more slowly, fading and degrading, as can be seen at the 20 minute point. After an hour, the dye is observed to be fully spread as a homogeneous mixture in the water. We can understand that the complex situation leads to the development of an emergent pattern, which is then subject to slow degradation and fading until the ultimate entropic mixing. All stages of the dye mixing are part of the entropic process, but there is an intermediate state of complex pattern on approach to the perfectly mixed state.

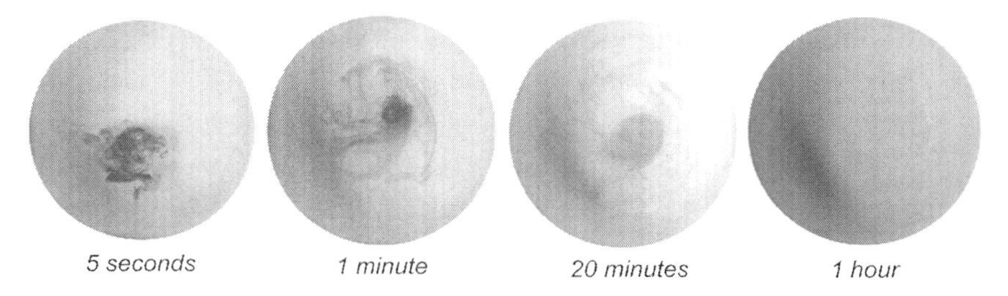


*Figure 6.17*

Due to the hierarchical structure of nature's systems, an individual organism can be seen, by increasing the perspective of scale, as a collective system of cell-systems. The individuated cells that make up the organism can be seen to possess emergent energy, with the emergent state representing the differentiation state of the cell.

The life cycle of any system-of systems can be understood to be a progression from low to high emergature/entropy under the influence of an emergent heat source, leading to redistributions of system elements in their emergent energy states (Figure 6.18). At birth, or in any organic beginning, there are large numbers of early stage undifferentiated states. This is the seed, egg or grassy field — which is primarily comprised of individual units of one unspecialized type. In this seed, egg, or grassy field, the positional states are clustered or spread in a highly symmetrical configuration such as a sphere. This beginning system, like the initial drop of dye in the water, represents a state of very low emergature and entropy, due to its primary occupation of low emergent energy states, and highly symmetric positional configuration.

If relevant emergent heat sources are applied, and the system is sufficiently 'insulated', growth, with increase in complexity, will begin. In analogy to the temporary pattern formed when the dye mixed in water, the complex relationships and interactions between elements generate a temporary emergent pattern state, which is the organization of the biological form. It is hypothesized that this *emergent pattern* phase of the system-of-systems, represented by some intermediate distributions of the system elements, optimally supports an emergent energy (such as the life of the organism). However, as emergent heat continues to be applied, emergature and entropy continue to increase, and the distribution of the system continues to disperse, leading away from this 'optimal' emergent pattern state. This spontaneous movement away from the emergent pattern is what we observe as aging. In death and decay of the physical form, all emergent energy and positional states become equally possible. There is therefore no longer an emergent pattern, and correspondingly, there is no longer support for the emergent energy of the system-of-systems. This is analogous to the state of the dye completely mixed within the bowl of water.

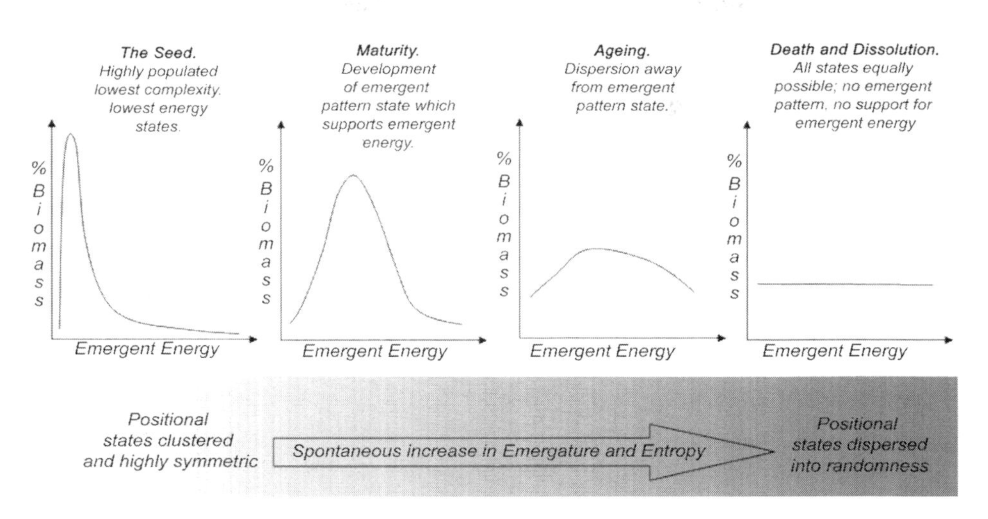


*Figure 6.18*

In terms of thermodynamics, these spontaneous progressions of life-cycle in a system-of-systems appear as a change in the spread of occupied energy states, with sharp, low energy distribution in the beginning, and a wide, homogeneous spread through all possible states in the end.

There is of course one more element to the life cycle. Not only can a system-of-systems be seen to increase the *dispersion* of system elements in emergent energy states (to increase emergature), but they often also manifest a great deal of form (biomass) very quickly and spontaneously in the beginning of the life cycle. This other spontaneous progression can be seen to result from another universal tendency which strives to minimize potential energy. In this (more loose and leafy) conceptualization, the seed can be seen as a potential emergent energy, akin to a ball at the peak of a potential energy landscape, which is driven to spontaneously fall into the well and manifest as organic form (emergent energy) given the slightest perturbation (such as soaking the seed in water). The amount of form produced is related to both the quality of the heat influx and also the depth of the information potential energy well. This perspective is shown in Figure 6.15.

## *A great conceptual fractal*

*The Universe is built on a plan, the profound symmetry of which is somehow present in the inner structure of our intellect.*
~ **Paul Valéry**

A fractal is a geometric object with a particularly complex appearance. Pondering it in its entirety may even hurt one's head. Yet, the magnificent thing about a fractal is that it is usually easy to understand it as a hierarchical structure created by iteratively executing an algorithm on a simple set of starting elements. Understanding the simple creation process of a fractal allows us to understand the nature of the complex whole.

For instance, as seen earlier in the Sierpinski Triangle, we have the simple transition (Figure 6.19), which is to be repeated wherever an upright triangle is found in the evolving form.

Perhaps because of the mind's tendency to understand the 'unknown' in terms of the 'known', a fractal is itself a good symbol for our very conceptual systems. This is to say that examining our interpretations of reality, and the ideas of organic mechanics and emergent thermodynamics, there appears to be a hierarchy (or holonarchy) of interconnected conceptual systems exhibiting self-similarity at different levels of conceptual perspective. While, as in the case of the geometric fractal, it may hurt our heads to consider this 'conceptual fractal' in its entirety, the conceptual fractal can be reduced to a set of 'simple' elements with a corresponding algorithm relating these components. The advantage

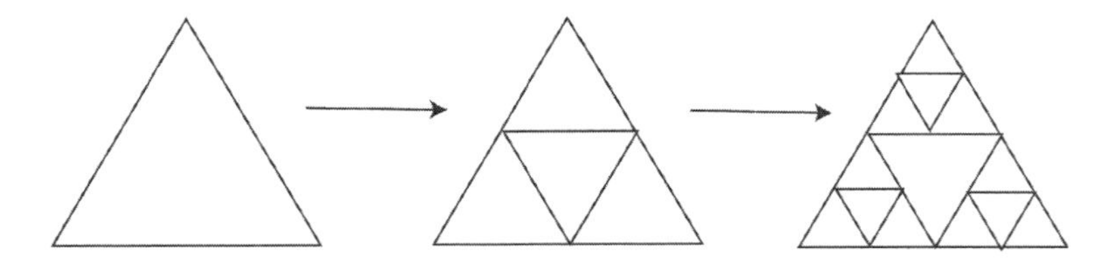

*Figure 6.19*

of recognizing this apparent structuring of reality is, hopefully, to recognize our inherent capacity for imaginative rationality, in which we can actively, creatively and appropriately envisage and select conceptual metaphors for the comprehension of unknown phenomenon. This is an alternative to reductionalist approaches to understanding. This is what has been done in the inventions of organic mechanics and emergent thermodynamics.

In our walk through the mind garden we have seen how human beings have developed conceptual systems that are able to effectively structure our experiences (observations) with the material universe at different levels of scale perspective. These scale perspectives are coarsely divided into: the microscopic realm of atoms, electrons and nuclei, understood using quantum mechanics; and the macroscopic realm, with relevance to everyday embodied human experience, which can be understood using thermodynamics.

An aim of the concepts presented in this book is to understand the supra-macroscopic behaviour of a collective system of living macroscopic subsystems. Organic mechanics determines the nature of emergent energy states of living things, using parallels to quantum theory. Emergent thermodynamics maintains that collectives of living things, representing occupied emergent energy levels, become directly analogous to a closed thermodynamic system which can reach thermodynamic equilibrium.

However, there is a plurality of these concepts which may give them wider applicability. For instance, an individual cell can be recognized as the system element which occupies an emergent energy state (emergent cell life energy reflected as the differentiation state of the cell), while the mature biological form, a collective of cells, is seen as the emergent system-of-systems. Simultaneously, the mature plant form can be seen as a system element which occupies a discrete state of emergent energy, and a collective of plants can be seen as the emergent system-of-systems. Furthermore, embedded sub-hierarchies of emergent systems are apparent. For instance, the organization of neurons in the brain can be seen to support various states of being, which can also be seen as an entirely new form of emergent energy (an energy of thoughts, awareness, emotion, and so on). Each system and system-of-systems can be contextualized using concepts from organic mechanics, or emergent thermodynamics, respectively. This is what is meant by a great conceptual fractal.

By focussing on only one portion of the great conceptual fractal at a time, each level can be seen to reflect the characteristics of the whole. Each part of the emergent system hierarchy can be seen alternatively as a microscopic system, or macroscopic system-of-systems, where the microscopic and macroscopic systems can be contextualized using similar conceptual protocol (organic mechanics and emergent thermodynamics, respectively).

The utility of defining a great conceptual fractal as such is that the same basic structuring pictures and fundamental laws can be seen to apply to a great number of phenomena, with the potential for practical applications in agriculture, ecology, medicine, and psychology/personal development. These practical applications will be touched on in Chapter 7.

We have seen that from the perspective of emergent energy and emergent closed systems, the progressions of living systems from states of simplicity to intricate complexity occur as spontaneous processes to maximize their entropy. Moreover, by merely assuming the existence of emergent energy states (life energy) we are able to account for some of the spontaneous processes of life and things living. We have effectively deduced the existence of life-energy as a real-world thing with real-world effects.

# 7. Harmoniously Human

*And the cost of a thing it will be remembered is the amount of life it requires to be exchanged for it.*
**~ Henry David Thoreau**

## *Addiction to ancient sunlight: a dire dependency*

Modern civilizations around the globe have an addiction to ancient sunlight in the form of fossil fuels (coal, petroleum/oil and natural gas), which are the highly concentrated forms of ancient sun-eating plants. We have built our great kingdoms, and the infrastructure of these kingdoms, around this ancient sunlight as a widely available, cheap energy source. Most obviously, fossil fuels provide energy for the transportation of people and goods. The approximately 600 million cars driven each day on planet Earth, as well as buses, trains, planes, and cargo ships are almost all fuelled by refined petroleum products. Fossil fuels also provide the majority (83%) of the world's electricity.[1] Electricity underlies the very basis of our modern lives, powering the lights, televisions, computers, stoves, water pumps, air conditioners, and heaters.

What's perhaps less realized is that the food that we eat — the entire agricultural industry — is highly dependent on cheap sources of fossil fuels. The growing of food in the developed world has become a centralized, heavily mechanized industry involving monstrously powerful entities such as large tractors, grain threshers, and land tillers, which are of course all fuelled by petroleum products. It's estimated that this industrial agriculture uses 10 units of fossil fuel energy to produce just

1 unit of food energy! Moreover, industrialized agriculture, which needs to force the land to produce as much food as possible due to its centralized nature, relies heavily on chemical fertilizers to grow food — chemical fertilizers that are actually made out of petroleum. And where does your food *come* from? In your local grocery store you'll likely find pineapples from the Philippines, mandarin oranges from Italy, grapes from Chile, and apples from New Zealand. Many grocery store items have been flown or shipped to your country, often from distances of several thousand kilometers, on planes or ships consuming fossil fuels. They are then loaded onto transport trucks, where even more fossil fuels are consumed to bring the foods to your specific town. If you live within a large city, just consider what you'd do if the networks bringing you water, food, and electricity were disrupted.

Finally, the physical material of our human world makes heavy use of fossil fuels that have been converted into plastics and asphalt. The roads that cars and trucks drive upon (spanning over one hundred million acres of land worldwide) are prepared from a black, sticky, substance (asphalt) extracted from crude petroleum. The plastic bags, spatulas, spoons, cartons, and drinking glasses that are involved in daily activities are all prepared out of converted fossil fuel material.

In its current setup, our basic needs, transportation, food, industry, and economy would collapse if the supply of fossil fuels were to dwindle or even become less affordable. What's particularly distressing is that the development of all of these innovative infrastructures powered by fossil fuels has facilitated a boom in the world's human population from approximately 1 billion prior to the major industrial revolution of the 1850s to nearly 7 billion in only 160 years, which is roughly seven times in only two to three human generations. Today we are dependent on the god-like, mountain-moving power stored in fossil fuels to support the infrastructure which supports our tremendous population. Petroleum is effectively the blood of modern human civilization.

Yet fossil fuels are a non-renewable energy source, and at the present rate of consumption, it's estimated that there are but fifty years' supply of petroleum remaining in the Earth's crust. Some people are content with this number, believing that the effects of dwindling petroleum will be sufficiently delayed to allow new technologies to replace them.

However, there are a few serious problems. First of all, fossil fuels require energy to be pumped from their reserves deep in the ground. Also, we are not only dependent on fossil fuels, but on a widely available, inexpensive source of fossil fuel. Finally, there is no viable, renewable alternative energy source that will meet the needs of our present situation.

A basic tenet of economics is that of supply and demand. If a product is widely available and there is a good demand for it, its price will be much lower than that of a product that is in scarce supply and high demand. Until recently, the amount of oil pumped out of the ground has exceeded or met the demand of society. However, as the amount of oil in the ground diminishes, it is more difficult to extract it from the ground. Therefore, there is less possible oil that can be produced by the oil pumps (less supply), and it also costs more in terms of energy to pump that oil out of the ground. This means that the cost of oil goes up — dramatically. As the amount of oil in the ground gets lower, the amount of energy to pump it out of the ground gets higher. Eventually more energy (in the form of fossil fuels!) is burned extracting the oil than is returned by the energy of the oil itself. This happens somewhere after the half-filled point of an oil well — and long before the ground is dry of oil. This whole case-scenario is known as *Peak Oil Theory* and has been corroborated by plenty of evidence from the behaviour of real oil wells. According to Peak Oil theorists (and including the highly optimistic estimates of large oil companies in an averaged estimate), the world oil supply is expected to peak somewhere around 2013, after which we must say goodbye to the era of widely available, cheap energy. According to Peak Oil theory, our society is already a runaway train headed for the edge of a cliff with tremendous momentum and very little action being taken to stop, plan, or even acknowledge this situation.

But there are alternative fuel sources, right? We've likely heard of *biofuels* produced from corn and other plants and *biodiesels* produced from the fat of animals. For the case of alternative fuels, there just isn't enough land mass on Earth, or even water in the ocean, to grow enough plant material to even begin to meet our present levels of consumption.[2] The world's current consumption of oil in one year is equivalent to more than four hundred times the amount of all the plant matter that grows on the Earth in one year, including the vast amount of plant material in the oceans. Each gallon of gasoline represents 98 tonnes of original plant material! In North America, each person consumes, on average, 25 barrels of oil each year. This represents the consumption of 102,900 tonnes of plant material per year, per person! To grow this much plant material would require approximately 42,000 acres of viable land, *per person*. Looking at oil this way shows us what a tremendously rich energy source it is, and also, how difficult, perhaps even impossible, it is to replace it to meet our current energy needs with a renewable, grow-able source.

But there are alternative energy sources such as nuclear power stations, windmills and solar power, right? While nuclear power represents a truly viable source of energy, it is risky, produces highly toxic wastes that remain seriously dangerous for hundreds of millions of years, and is based on uranium, another non-renewable material destined to be used up within the next forty years. At present we base 83% of our electricity generation on fossil fuels, and there's a reason for that — they represent an immensely concentrated source of energy compared to the energy put into obtain and maintain them. Wind farms and solar power require large, expensive apparatus requiring manufacture and maintenance. Energy certainly comes out of them, but at a much weaker level when the amount of energy put into their manufacture and maintenance is accounted for.

In short, substantial evidence suggests that while we will be able to produce *some* energy sustainably, there is no way that we will be able to meet our current energy uses by these alternative sources. Society as we know it is perched on a precipice of dramatic change. We're faced with a growing population representing many more mouths to feed, which will require more resources, and we're going to have to do this with much less resources available — resources we've become highly dependent on.

Even if you don't happen to believe in Peak Oil Theory, fossil fuels are just one of a number of factors that indicate humanity is indeed a runaway train heading off the tracks.[3] The burning of all this fossil fuel has released a tremendous amount of carbon dioxide into the atmosphere, leading to a warming up of the entire Earth and an unpredictable climate featuring record heat, record cold, record droughts, record floods and record hurricanes. These unpredictable and harsh weather conditions will make agriculture even more difficult. The planet-regulating, oxygen-producing, carbon dioxide-fixing Amazon rainforest is being destroyed at a rate of 200,000 acres per day! Thirteen million tonnes of toxic chemicals are released to the environment every day. Each year 25 million acres of arable land is transformed to infertile desert conditions due to unsustainable agricultural practices and the growth of concrete and asphalt paved urban environments. As a consequence of the above phenomena, one hundred and thirty plant and animal species are driven to extinction *each day* while in that same day the human population grows by over 200,000. In the oceans 90% of the big fish (tuna, marlin, sharks, dolphins) have been killed by humanity's intensive commercial fishing operations, leaving only a meagre number remaining.

In the very near future, it seems very likely that we will need to do more with less, but what and how? Without being able to heavily rely on the brute-force strategies fuelled by machines powered by fossil fuels, it is a more thorough, functional, practical understanding of the dynamics of natural ecosystems which will become more valuable now then they've ever been. This is true for the rehabilitation of deforested and mined landscapes, and the growing of food itself.

## *A toolbox of techniques from emergent thermodynamics*

Gaia theory envisages the entire planet Earth as a living organism with interconnected holonic 'organ', 'tissue', and 'cell' systems. The Gaia metaphor cultivates an entirely different relationship between human beings and the Earth, allowing for a stark re-interpretation of our human selves in relation to the dynamic, highly regulated and interconnected environment in which we are embedded. We are distinctly cells in Gaia's body, and like a body she is clearly our supporting system. We depend on her; she does not depend on us. In the context of our self-assembled, exergy degrading, petroleum-fuelled concrete kingdoms we are behaving no differently than an autoimmune disease, cancer, or spreading fungus on a peach, destroying the host supporting system of the body in which we are a part.

More specifically, the notion of Earth-as-organism is the overriding conceptual metaphor in a complex systems theory allowing us to identify and relate the many interconnected parts of the Earth to account for the delicately balanced, life-supporting whole. From this perspective, human activities are interfering with Gaia's regulatory processes in terms of her various primary organs: the atmosphere, hydrosphere, soil, and most prominently the biosphere. Nearly every activity that humanity has embraced interferes with Gaia's regulatory mechanisms and disrupts the delicate balance of the body that is Earth. As discussed previously, plants (namely mature forests) are the most influential organ systems in the body of Gaia, as they regulate carbon dioxide and oxygen concentrations in the atmosphere; play a key role in regulating the hydrosphere by drawing liquid water from the atmosphere to provide water for themselves and drinking water for other creatures; create new soil in the decomposition of their shedding plant parts; provide food and resources for other living creatures; and cool the temperature of the Earth by acting as sunlight exergy degraders and carbon dioxide sinks. From the perspective of Gaia theory, it is plant life which has taken over this once barren rock floating in space, and transformed it into a vibrant, regulated living thing capable of supporting even more life and complexity. Yet we are destroying forests and arable land at an unprecedented rate.

While the Gaia Theory paints our relationship to the Earth in beautiful broad strokes and wide-scale perspective, emergent thermodynamics helps by providing a framework of more specific details as to how we might go about aligning human interests with the dynamics of the greater systems in which we are embedded. Emergent thermodynamics yields simple, straightforward, and elegant insight into the movements of whole plant communities by short-circuiting the complexity of complex systems science perspectives. Systems science perspectives still focus on the many parts of the system, and the specific interactions between those parts, in order to account for the properties of the whole. In contrast, emergent thermodynamics works with those parts in an entirely new context (the emergent system). The focus of emergent thermodynamics is on the processes and evolutions occurring in the new context of the whole emergent system, which yield generalized results that may make the exceedingly detailed and complex scenario manageable. Let's take a moment to review emergent thermodynamics and in the next section to consider how it can structure more harmonious human activities in a post-fossil fuel world.

In its present stage of development, emergent thermodynamics focusses on plants and collective systems of plants (plant communities and ecosystems), which as we've just discussed, are the most influential organ systems in regulating the life-supporting Earth body of Gaia. In Organic Mechanics we saw that each plant could be associated with morphogenetic wavestates and a corresponding emergent energy (emergent energy) state. By looking at the forms of many different kinds of plants we were able to see archetypal repetitions in their patterns that suggested plants belonged to a bigger system of possible wavestates. In emergent thermodynamics we see each plant as occupying a particular morphogenetic wavestate in the great morphogenetic wavefunction of all possible plant forms, and to thereby have a specific amount of emergent energy associated with it. With this kind of conceptual set up, it's possible to take a look at any old plot of land, see what kinds of plants are growing there, find out what morphogenetic wavestates and emergent energy each plant represents, and show the distribution of plant matter in terms of these various emergent energy states (as an emergent energy state density diagram). We can look at the distribution of plants in their emergent energy states and based on that distribution, define an emergent temperature (emergature) for the plot of land. In doing this, we are simply following how temperature is defined for a spatial region of molecules distributed in kinetic energy states. In examining different regions of land (the grassland, meadow and woodland) we've observed that the distribution of plants in their emergent energy states as a function of time behaved similarly to the distribution of molecules in their kinetic energy states when exposed to a heat source. The grassland, with a narrow distribution in low emergent energy states represented a low emergature. The deciduous woodland, with a wide distribution through many emergent energy states from high to low represented a high emergature. Under the influence of emergent heat sources, the bare land spontaneously 'warms up' from low to high emergature, from grassland to woodland. Emergent heat sources for plants represent water, sunlight, carbon dioxide, and physical temperature. We proposed that the final emergature of plants on the landscape depends on the intensity of emergent heat sources in a particular region, with a colder, dryer climate like the Canadian shield producing a lower emergature final plant community than the higher emergent heat region of the Amazon rainforest. A place like Antarctica receives amongst the lowest possible emergent heat flows on Earth, which is reflected in its essentially absent plant communities on its landmass. Finally, we defined the concept of insulation for emergent systems. The plant community needs to be 'insulated' in order to advance to higher emergatures. In emergent thermodynamics insulation essentially means that the wastes of the emergent system are actually a resource as well. For instance, dead and discarded plant material represents an organic fertilizer (compost) that assists in the growth of young plants.

Emergent thermodynamics provides an entirely new context that can assist us in accounting for the natural tendencies of collective plant systems. Emergent thermodynamics can therefore provide practical tenets that indicate how we humans can align our interests with the natural tendencies of natural systems, instead of working hard to fight against them. Emergent thermodynamics applies most readily to agriculture in an age without ample fossil fuel fed machines, and to land rehabilitation practices.

The following seven practical concepts can be drawn from the concepts of emergent thermodynamics. As we'll explore in the following section, these practical principles can assist in focussing and aligning human self-interests and activities with the natural dynamics of plant communities:

### *The emergo-dynamic Earth model*

In the emergo-dynamic model of the Earth, emergature is analogous to temperature (see Figure 7.1 to help get the concept). However, emergature represents something very different to physical temperature.

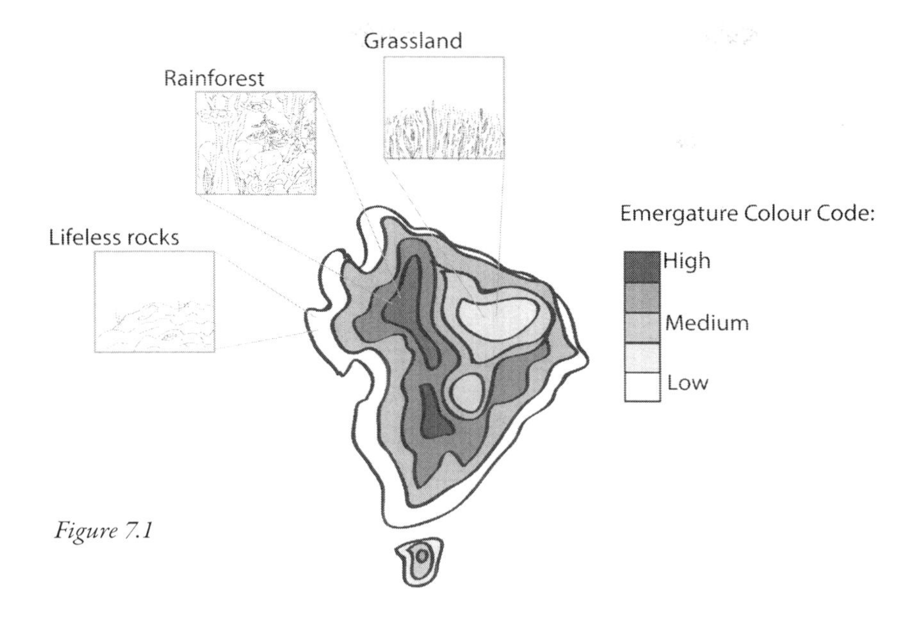


*Figure 7.1*

Emergature represents the developmental stage of a community of plants. We imagine that the land masses of the Earth are like sheets of metal receiving a heat source. In the analogy, the land masses, just like sheets of metal, 'heat up' in particular regions due to the heat source. In 'heating up' it is the emergature of these regions that increases. This signifies the development of bare land to mature plant communities. If some event occurs, such as the eruption of a volcano that completely kills all plant life and covers the land in stony, hardened lava, the emergature of this area has been reset to zero and it represents a 'cold spot' on the metal sheet.

*Emergature estimates*

The emergent thermodynamics model is connected to real life situations as we can literally take observations of the type of plants growing in any particular region to estimate the emergature of that region. We do this by constructing a state diagram that plots all of the plants in the region according to the morphogenetic wavestates/emergent energy states implied by their form (you can revisit this idea in Chapter 6 Figures 6.5, 6.6, and 6.7, pages 195–202). The equilibrium emergature of a region can be estimated by investigating the state-density diagrams of mature plant communities (100–200 years old or greater) growing in the region. This is analogous to how a single temperature is defined by examining the state density diagrams of distributions of molecules in their kinetic energy states.

### *The life-cycle-impetus*

Emergent thermodynamics proposes that all emergent systems under the influence of an emergent heat source are driven to spontaneously proceed through to the equilibrium emergature of the region. This spontaneous drive can be likened to a force of nature with tremendous impetus to move forwards — like flowing water constrained at the top of a hill. Moreover, if the system is at low emergature, it cannot spontaneously jump to a higher emergature, but must progress through the characteristic life-cycle. The life cycle begins with plants distributed in low emergent energy states and proceeds to plant forms distributed throughout a wide range of low to high emergent energy states. This is simply saying that the emergent system is following the same universal law of entropy that applies to other non-living collective systems. This necessitates that states of low energy (simplicity) be filled first, and only with continued heating in an insulated (self-sustaining) system, will occupation of higher states become possible. This explains why we do not spontaneously observe a 'plague' of high emergent energy species such as orchids and oak trees growing in a deforested or devastated region. Instead we see lichens, algae, grass, and other low emergent energy species.

### *Active freezing and refrigeration*

The clear-cutting of an area is akin to creating a cold-spot of near zero emergature. The paving over of land is akin to freezing it at this near zero emergature. The use of herbicides is akin to refrigeration of the land at a low emergature as it represses the evolution of the system to later stages of its life cycle.

### *Active heating*

Adding chemical fertilizer to a region is akin to increasing the emergent heat influx to the region. This addition of emergent heat drives the system to progress further into its life cycle.

### *Emergature flows*

The emergatures of regions of land behave like hot and cold spots of a metal sheet, which redistribute the hot or cold spot to an average temperature between the two. A small low emergature region surrounded by lots of high emergature land is akin to a small cold spot surrounded by a large heat sink (top image of Figure 7.2). The small low emergature region will quickly warm. In contrast, a small region of high emergature land surrounded by low emergature land will not be self-supporting as the emergature will decrease to due to the large surrounding cold region (see bottom image of Figure 7.2). This reflects the collapse of small, isolated pockets of high emergature land. For instance, if a small region of the Amazon rainforest were left standing with only open, barren fields surrounding it, it dies. From the system-science perspective this occurs because the Amazon rainforest produces its own water supply and requires a large region intact in order for it to sustain itself. From the emergent thermodynamics perspective this occurs because the small pocket of high emergature land will lose its emergature to the large cold region around it.

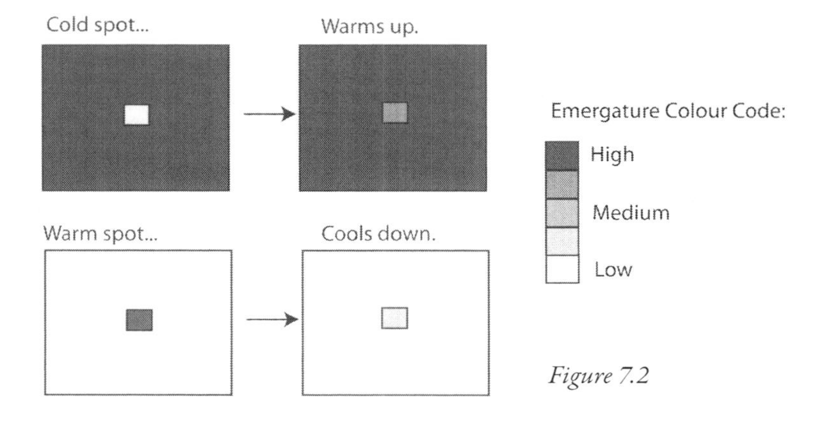


*Figure 7.2*

*Priceless value of high emergature regions*

A high emergature region of land that's at or near its equilibrium emergature is a priceless resource to humanity due to its stability in the face of climate change, potential to produce valuable items for humans and other species, its ability to act as an emergature sink (see concept 6), and the fact that they require many human generations to develop!

Let's keep these seven principles in mind as we discuss the current state of, and a plausible, harmonious way forward, in humanity's food-growing enterprises in a post peak oil world.

## *High emergature agriculture*

The aspiration of a sustainable humanity could be defined as the self-interest invested devotion to life and *all* things living. Human beings have demonstrated exceptional creativity and adaptability; however, by placing an immensely high focus on the gratification of explicitly human interests, and valuing other species only in terms of their immediate human usefulness (which usually means dead!), we have become separated from our supporting system (the rest of life). We've used our creativity, ingenuity and adaptability to pursue self-interests isolated from the natural world; self-interests that see nature as either a material resource or a hindrance to be conquered. Consequentially, we continue to actively work against nature or to simply eradicate nature in favour of fulfilling our short-term interests. Cultural change towards sustainability requires widespread individual change of heart between each of us humans in relation to our body Earth, as well as practical concepts and steps to embrace as new means to fulfilling the required ends of humanity. In its current state of development, emergent thermodynamics applies most readily to one of the most important and practical things we're going to need to do as a species: grow food, medicine and resources without using brute-force methods powered by fossil fuels. Let's begin by taking a look at the current state of agriculture from the vantage of emergent thermodynamics.

The current state of 'mono-crop' agriculture (that is, one crop agriculture) typically begins with a clearing of high emergature land in order to grow and maintain a comparatively low emergature region as crops. In the Amazon rainforest, this is occurring right now as very high emergature land is cleared to grow simple mono-crops and grass for cattle. In New Zealand this extensive deforestation occurred in the mid 1800s when newly arriving European pioneers cleared 87% of the existing high emergature equilibrium forests in order to grow mono-crops such as wheat, or grass for sheep and cattle. After this initial clearing, farmers may attempt to maintain a very low emergature situation, typically consisting of a mono-crop of a plant such as wheat. Wheat, and other cereal crops (barley, oats, rye, rice) are a type of grass with two levels of structural hierarchy (organization of the leaves and organization of the leaves on stem) requiring the more simple blade and fishbone fields to account for their structure at the leaf and leaf-on-stem levels, respectively. Thus, cereal crops have a low complexity number/emergent energy value of around 32. Other crops, such as legumes (soybean, mung bean) also have two structural levels of hierarchy, with a slightly more complex leaf structure accounted for by the fishbone field, giving them a complexity number/emergent energy value at a slightly higher 33. To grow only a cereal crop *en masse* in a particular region would be to have a large amount of plant forms confined to the low emergent energy value of 32, which, as we saw in the example of the grassland, represents the first stage of a region's development. Moreover, it may represents a dramatic drop in the emergature of the region away from the equilibrium emergature.

Some regions on planet Earth, such as the African savannah, the extensive prairies of the North American Midwest, and the steppes of Europe (Russia) have equilibrium emergatures which are precisely that of grassland. In these regions, most plant material remains concentrated in the low emergent energy states. A low equilibrium emergature of grassland is consistently correlated to a low emergent heat source for the region, which is typically a lack of rainfall. However, for many other areas of planet Earth the equilibrium emergatures of the region are much higher than grassland. This means that once cleared, the land will have an impetus to proceed through its life cycle to reach its equilibrium emergature. Considering a field of a mono-crop, the impetus to 'warm-up' may show up as the presence of weeds and other undesirable species such as insects. The farmer may then confuse the situation by using herbicides and pesticides to kill off the newly emerging species, which represent the system's natural attempt to move towards its equilibrium emergature. At the same time more emergent heat, in the form of fertilizer, might be added to the crop in order to increase the yield of the low emergent energy mono-crop. This further increases the impetus of the system to move towards more advanced stages of its life-cycle. From the view of emergent thermodynamics, mono-crop agriculture in regions with a much higher equilibrium emergature than the emergature of the mono-cropped region is an unnatural situation, equivalent to refrigeration, which requires us to actively invest energy in an attempt to prevent the community from 'warming' up to the fully dispersed occupation of states.

In addition to the above concerns, in an era without brute force strategies powered by fossil fuels mono-crop agriculture will become dramatically less efficient in a time when there will be more mouths to feed and efficiency will be paramount. Moreover, mono-crop agriculture of annual plants that require the soil be re-tilled, new seeds planted, and plants to grow each year is very hard on the environment and is very energy intensive. A viable alternative to mono-crop agriculture is to create, however simply, a real community of plants involving a number of elements that are self-supporting and grow in complexity and productivity each year. All of the plant elements should also have some utility for human beings, whether these are nutritional, medicinal, or structural properties. This alternative strategy to mono-crop agriculture has been called permanent agriculture (permaculture). Permaculture is a practical philosophy for human land use which is designed for the dual purposes of meeting human needs abundantly, while also acting to steward natural systems, encouraging them to grow and thrive in full complexity and vitality. In the words of one of its founders, Bill Mollison:

> Permaculture is the conscious design and maintenance of agriculturally productive ecosystems which have the diversity,

> stability and resilience of natural ecosystems. It is the harmonious integration of landscape and people, providing their food, energy, shelter and other material and non-material needs in a sustainable way.
>
> The philosophy behind Permaculture is one of working with, rather than against nature, of pro-acted and thoughtful observation rather than protracted and thoughtless action; of looking at systems in all of their functions rather than asking only one yield of them, and of allowing systems to demonstrate their own evolutions.[4]

An emergent thermodynamics perspective can assist in the design and development of food producing plant communities as agricultural enterprises for human beings through its basic concepts. First of all, we can estimate the equilibrium emergature of the particular region we wish to grow food in by examining the state density diagrams of mature (one hundred years or greater) plant communities in the area. The goal is to eventually develop a permaculture enterprise — a food and resource producing plant community — that is geared towards the equilibrium emergature of the area. Thus, if we are trying to grow food on the Russian steppes where the equilibrium emergature is quite low, we will aim to grow a perennial food producing grassland. However, if we are in a tropical region that supports rainforest, we will aim to grow a complex forest garden including a wide variety of species from low to high emergatures. If we are working with land whose emergature has been zeroed (deforestation, natural disaster, and so on), emergent thermodynamics posits that we will get the best results if we begin our permaculture enterprise at low emergatures and actively promote progression of the life-cycle to higher emergatures, aiming for the equilibrium emergature of the region. Now what exactly does this entail?

Emergent thermodynamics shows us the relative proportion of various archetypal plant species at different levels of emergent energy during the natural life-cycle of a plant community. So for instance, we've seen that a field of grass will spontaneously evolve into a distribution of grass, milkweed, sumac, birch, and oak over time (top panel of

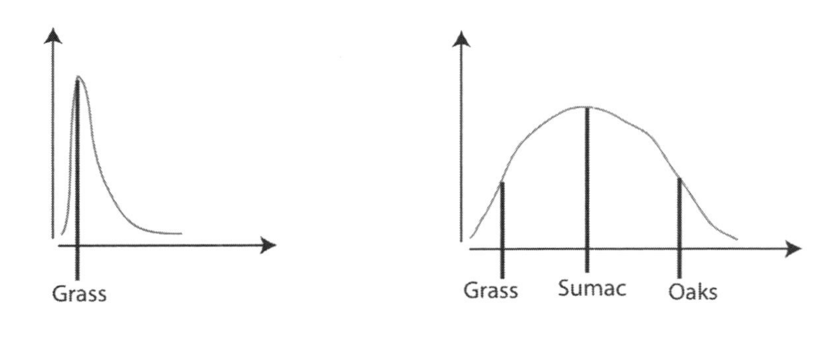


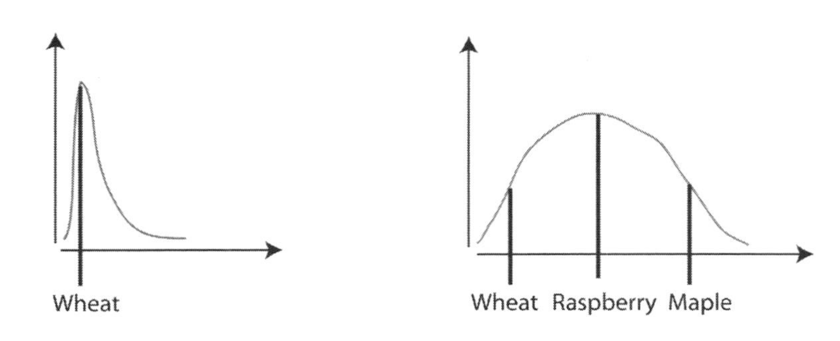


*Figure 7.3*

Figure 7.3). The main idea of emergent thermodynamics in permaculture is that as emergent energy states are archetypal, there are many different plant possibilities available at each emergent energy level. Thus, we can engineer a harmonious plant community by selecting plant elements that are of some benefit to humans but which also comply with the distribution required by the stage of the plant community's life cycle (bottom panel of Figure 7.3). Instead of grass, we can begin by growing a food producing grass such as perennial wheat. In aiming to steward the progression of the plant community's life-cycle we actively introduce similar emergent energy species to those that would occur naturally, but select ones that produce human foods or resources. Thus, perhaps we'd introduce raspberry in place of sumac, apple in place of birch, and syrup-producing maple trees instead of oaks.

In order to fully apply emergent thermodynamics for construction of food producing plant communities, we must continue to study the distribution of natural plant communities (in emergent energy state density diagrams) at various stages of their development. We need to be able to parameterize the distributions that we obtain from these observations. It is hypothesized that these distributions will have a consistent nature independent of the specific species that are present, and dependent on the age of the plant community and the amount of emergent heat in the region which dictates the equilibrium emergature. At present our agricultural activities fight against the system's tendency to progress to its equilibrium emergature. Human beings must strive to develop insulated emergent systems that are near equilibrium emergature as agricultural enterprises to provide food, medicine and shelter in the self-interest of humanity. Unfortunately, none of this is a quick-fix solution, but requires a long-range view and cultivation spanning a few human generations.

## *Reflections on human activities and the environment*

While emergatures are leveled to initiate mono-crop agriculture, humans write themselves out of their agricultural systems and carelessly supply more emergent heat to alternative regions. For instance, human sewage, representing an emergent heat source for plants and produced in vast quantities by our mushrooming population, is largely dumped into oceans and other waterways, when it could be safely composted and used to fertilize agricultural enterprises. Instead, after we throw away a natural source of fertilizer we then use extensive chemical fertilization that dramatically increases the level of phosphate and nitrogen in the soil and water. Our use of fossil fuels increases the level of carbon dioxide in the air and water, increasing the physical temperature of the Earth and the amount of available carbon dioxide and moisture in the environment. In conjunction with this 'leaking' of our wastes into the larger environment, human activities tend to zero the emergature of vast regions of the Earth. Emergature is lost in activities such as the clear-cutting of forests, the paving over of land masses with asphalt to erect vast savannahs of concrete called cities, and in commonplace activities such as the maintenance of well groomed, homogeneous, weed-free lawns. The combination of zeroed emergature and high emergent heat 'leakage' is a formula encouraging the spontaneous development of low emergature eco-systems which rapidly attain the necessary biomass to match the 'energy' existing as available emergent heat. This is the case as, if emergent energy states exist, states of low energy (simplicity) must be filled first, and only with continued heating in an insulated (self-sustaining) system, will occupation of higher states become possible. Our activities favour 'foggy' blooms of low emergent energy states that represent such phenomena as overgrowths of algae, fungus, micro-organisms, weeds and certain insects, all of which represent the first-phase of organic development — the occupation of the first rungs of the emergent energy 'wavefunction'. By continuously reducing the emergature of living systems on Earth while supplying extraneous emergent heat, we are generating a situation which guarantees a tremendous insurgence of simple, weedy organisms, which we otherwise would call plagues.

We have an interesting response to the manifestation of plagues. According to emergent thermodynamics, a solution to plagues focusses on developing strategies to encourage the evolution of the system through to higher emergatures representing more stable, diverse, and valuable ecosystems. Instead we have developed methods to eradicate the plague. Antibiotics, fungicides, herbicides, pesticides and methods of sheer physical force such as burning and clear-cutting are designed to accomplish these tasks. However, from the hypothesis of emergent thermodynamics, while effective in the short term, these methods only work against the natural impetus of the system-of-systems to progress to higher emergatures, treating the symptoms without understanding the cause. In addition, these methods of eradication can be (temporarily) very effective as well as being very non-specific, eradicating much more than the intended 'weedy species'. Introduced as toxins into ecosystems they serve to lower the emergature still further. The long term consequence of this strategy of mass eradication is an even stronger insurgence of the plague, as nature remains driven to spontaneously proceed through her entropy-dictated evolutions, and, after eradication, must do this beginning with a system with an even lower emergent energy and temperature.

Before the widespread use of fossil fuels in our great machines, and prior to the correlated mushrooming of our population, humanity's effects on ecosystems were typically limited to small regions surrounded by a large 'heat-sinks' of high temperature wilderness, which would readily re-populate any cleared region. This may have given us the illusion that with respect to the time-scale established by the average human lifetime, nature is a fast-healing, versatile system with the capacity to rapidly regenerate any 'wounds' created in her grand, wild vistas. However, in the past century, humanity's effects have, like the spread of a malignant tumour of asphalt, or a grey flair of concrete fungus, progressed quite extensively over the surface of the Earth, clearing, paving, constructing and otherwise 'developing' the land. In this day and age, wide expanses of high emergature wilderness no longer exist on most regions of the Earth. Consequentially, our actions have decreased the emergature of vast regions of Earth. This is a serious consequence, for though the Earth may have began as a barren rock surface with zero emergature and energy, the activity of sustainable (insulated) plant systems which have grown over thousands to hundreds of thousands of years has been required to develop a high biomass, complex ecosystem represented by a high emergature. This is a devastating legacy to a humanity whose lifespan is of the order of a hundred years, for it implies that we have created and are continuing to create irreparable and irreversible damage with respect to our children's, children's, children's generations.

## *Tuning the biological clock*

Emergent thermodynamics helps us to account for, and therefore, to interpret and work along with the life cycle of an emergent system-of-systems. While we've been focussed on plant systems, we've briefly discussed how the parts of the emergent system-of-systems may be interchangeable (Chapter 6, pages 218–220). Thus, while we've not developed the idea, a holonic entity such as a cell might be interchanged for a holonic entity such as a plant form. Working with this as an analogy, the emergent energy state of the cell is in relation to its differentiation state. Undifferentiated embryonic cells represent the lowest emergent energy states, while perhaps neurons represent the higher and most specialized emergent energy states. Consistent with the idea of life cycle developed in emergent thermodynamics, the developing organism progresses from a symmetric cluster of the same cells in the lowest emergent energy states to a complex organism representing a wide spread of cell types throughout many different emergent energy states. For an organism we could further postulate the existence of an ideal 'emergent pattern' state that represents the form of the mature organism, occurring at some intermediate point in the emergent system-of-system's life cycle. Further progression to later stages of the life cycle represents aging and death, which is movement away from the emergent pattern state which ideally supports emergent life energy. We discussed this in Chapter 6, pages 214–217.

This all occurs in response to emergent heat influx, which, as in the case for plants, represents flows of matter and energy required for growth of the organism. When emergent heat is input, an insulated (self-sustaining) emergent system spontaneously proceeds from a homogeneous, symmetric 'egg'-state representing the occupation of primary levels of emergent energy, to a complex, mature emergent pattern state that ideally supports emergent life energy. With continued 'heating', there is progression past this emergent pattern state with the more traditional signs of entropic increase involving denigration and dissolution of the emergent pattern. This is what we interpret to be aging.

If the organism is to survive, it requires emergent heat energy. Moreover, the organism is of course not a perfectly insulated system, as it dissipates energy in the form of physical heat and kinetic energy of movement, as well as waste matter. It requires a base supply of nutrients and materials for maintenance of physical structures and processes. The idea behind tuning the biological clock, however, considers that in a mature organism, emergent heat supplied in excess of essential requirements is extraneous to the system and can be used to 'heat' the system. This extraneous heating is very useful in the early stages of life, as it is allowing the organism to manifest the mature emergent pattern state. However, after maturation, this extraneous heat serves a rather counterproductive function as it will 'heat' the organism past the optimum emergent pattern state. The more extraneous heat is supplied, the more rapidly the emergature will increase. When applied past maturity, this extraneous heat represents accelerated aging and eventual death, as the dispersion of emergent energy states moves past the optimal mature state.

Remarkably, these predictions arising from emergent thermodynamics are substantiated by experimental evidence! Fasting, the voluntary abstinence from food, has been recognized for thousands of years as an immensely curative and spiritual practice. More recently, modern science has discovered that calorific restriction without nutritional deficiency; that is, eating slightly less or on par with the energy demands of the body without depriving the body of essential nutrients, is able to increase the lifespan of a number of organisms by up to a third, as well as slowing their characteristic signs of aging.[5] Rehashing the concept of the life-cycle in emergent thermodynamics, we can understand why this is the case. Extraneous heat (food energy) supplied after maturation accelerates the system-of-system's approach to entropic death. Therefore, the wise thing for any self-directing organism, or caregiver for organisms interested in slowing the aging and extending the lifetime of their life-forms, is to supply minimal extraneous heat to the organism once it has reached maturity. In plainer language this represents a low calorie but very high nutrition diet.

# Conclusions

*Every tree and plant in the meadow seemed to be dancing,*
*those which average eyes would see as fixed and still.*
**~ Rumi**

In the city, we are swimming in a sea of scientifically recognized energy. Car engines churn out mechanical energy, hot exhaust, and rumbling sound waves; televisions receive the electromagnetic (stuffless wave) fluctuations in the atmosphere and transmit sound and light wave energy to a watching observer's screen; neon signs, streetlights, headlights, and shop lights transmit further electromagnetic waves; transmission lines shuttle constrained electromagnetic waves through electronic conduits overhead and underfoot. The city is awash with kinetic energy, heat energy, and electromagnetic energy. All of this is the result of how we have been thinking, of what we have believed in, of what we have 'sung' into existence. However, some may perceive a great dullness in the city, a great emptiness, and some may feel the city to be an 'energy drain' on their body. This may be in part because the city is missing something integral to humanity's health and wellbeing — something found in abundance in natural environments such as forests and meadows.

In a forest, we are swimming in a sea of scientifically *unrecognized* energy. Imagine every tree embodies a field of energy. Every leaf, every frond, every shrub, every blade of grass is alive with energy. Every ant, every bumble bee, every cricket, every cicada, every weevil and maggot, every earthworm, caterpillar, firefly and monarch butterfly are alive with energy — kinetic, thermal, chemical, *and* life energy. We have seen here that the biological form of these living things may hold characteristic features corresponding to the quality and form of that associated energy. We have called this energy an emergent energy. This is a perspective I hope we come to recognize; one we make part of our rational realities, which is considered in our activities, big and small.

Yet we've seen in this book (Chapter 1) that science, in deciding for us what is rational, has become primarily focussed upon the three -isms thinking style in which living things are understood in terms of their reducible material parts engaged in cause and effect mechanisms to account for the behaviour of the whole. Despite all of the frenetic research activity of the life sciences, which accounts for roughly 80% of all scientific research today, there is rarely a mention or account of the property of life itself, and certainly little recognition of life as a unique form of energy.

Given the trust and allowance we've given science to define what constitutes 'rational' activities, the above view of living things means that life at all scales of perspective — from the tiniest single celled amoeba to the living Earth itself — are being objectified, devitalized, and dissected. When faced with a problem, most scientists are currently conditioned to seek the solution by asking questions such as: how can I reduce this to its material parts? What are the axiomatic factors at play here? How can I identify, isolate, extract and study them to determine their mechanism? The relationship between the scientist and their subject of study is one of objective emotional detachment. The consequences are arguably the cultivation of a narrow view incapable of considering networked influences in the larger scale of whole organisms and whole planets, resulting in very serious side effects and a strong emotional detachment inhibiting empathy. While holistic perspectives may offer substantial benefits, for circumstantial and political reasons, alternative holistic views focussing on the networked whole and emergent properties are rejected or simply ignored by the life science community. A change of the mainstream paradigm seems to be a very hard thing to sprout and grow in the defining majority.

We have seen (Chapters 1 and 2) that the problem with understanding living things by the three -isms style is that they exhibit strong emergent properties, and that those emergent properties have the ability to influence and dominate the behaviour of the very parts that created them (cyclic causality) as well as other causal properties acting on the world at large. In addition, the organized complexity of living things — their self-assembling intricately intertwined hierarchical structures and functions — make them particularly difficult to manage from a strictly bottom-up approach that focusses on their parts. New context and creative ingenuity are required for us to progress our understanding and ability to work with living things.

As we've explored in this book (Chapter 3), alternative thinking styles to the monopoly of the three -isms already do exist, and can be primarily found in the physical sciences. An imaginative rationality making use of a tool of the mind called conceptual metaphor is used to generate many axiomatic concepts of the physical sciences. In conceptual metaphor, the mind typically uses a modelling picture from a familiar situation to engender a new context in which to work with and understand an unfamiliar phenomenon. We've seen that in working with energy itself, the physical scientists employ metaphoric models such as that of a ball-on-a-hilly landscape or a flowing fluidic substance (recall heat and time are viewed as 'fluids') to work with the features of other systems that are related only in their abstracted characteristics to the model itself. Moreover, we've seen that in physics, simultaneous reductionist and imaginative rationalities are accepted as interpretations of the same phenomenon (for example heat as material molecules jiggling, and heat as a fluidic substance flowing between objects) and that each interpretation has utility for different situations.

In spite of the rejection or apathy they've received from the mainstream scientific community, the seeds of alternative thinking styles already exist in the life sciences. In this book we looked at the holistic ideas of life-as-energy, biological fields, and Earth-as-organism. While these three concepts have not been embraced by the mainstream sciences (to say the least) we've seen that they intuitively fit as functional conceptual metaphors. This makes them no different from various axiomatic ideas — potential energy, energy fields, the basic nature of energy itself, heat, or magnetism — many of which still do not have an account in terms of their reducible material parts. However, they remain widely accepted and used in the context supplied by their various structuring conceptual metaphors. There is absolutely no reason why the concepts of life-as-energy and fields in biological development cannot be taken seriously and explored scientifically in lieu of a three -isms explanation. On behalf of the above reasoning, in this book we decided to accept concepts like life-as-energy on behalf of their befitting underlying conceptual metaphor models, in spite of the current absence of a reductionist material based mechanism that could account for them. In deciding to take them seriously, we set out to explore and develop them and give them some substance. It was hoped that this exploration and development would lead to some novel systems for working with living things more accurately and with increased feasibility. This was also undertaken as a very long working example of a whole life science, where conceptual metaphors could be used to build new concepts and new context through an embrace of imaginative rationality.

In Organic Mechanics (Chapter 5) we developed morphogenetic fields capable of accounting for the patterns in leaves and plants as if they had guided the structural development. The mathematical form for these prospective 'fields' was developed, where no mathematical description had before been recognized for these patterns at an everyday world scale of human beings. The mathematics underlying the patterns was useful as it allowed us to precisely describe the proposed metaphoric fields and to generate images of what they might look like. We determined that there were a great many variations in leaf patterns, but only several archetypal patterns. We proposed that all of these pattern-archetypes belonged to a greater system and extended the conceptual metaphor to a whole new level by suggesting these were not merely fields, but could be seen as morphogenetic wavestates. We used quantum mechanics as an underlying model in a new conceptual metaphor. We further supposed that just like quantum wavestates, morphogenetic wavestates were associated with an energy. We called this energy associated with the biological form emergent energy.

We then went on to see what would happen in collectives of plants given their new description in terms of morphogenetic wavestates (Chapter 6). We worked with the plants growing in a region of land in the same way we'd work with atoms in a sheet of metal. We constructed state density diagrams, but in place of energy states we had emergent energy states. From the distribution or shape of the state density diagram we would normally be able to determine the value of the temperature of the collection of molecules. In this new case, from the distribution of the state density diagram we determined the value of the emergent temperature (emergature for short) of the collection of plants. Thus, for every region of land, by measuring the distribution of plants we could determine the emergature of the very region of land. This was analogous to determining the temperature of a spot on a macroscopic piece of matter (like a metal sheet). We discovered that just as a piece of matter will spontaneously heat up and increase in temperature when exposed to a heat source, that a barren plot of land will spontaneously increase its emergature in response to an emergent heat source. The change of the type of plants growing on a landscape with time is entirely consistent with the change in temperature of an object. It begins with a high concentration in the lowest energy states and proceeds to a wide distribution across all emergent energy states, high and low. This allowed us to account for the life cycle of a plot of land in terms of the age-old rigorously tested thermodynamics principles representing a whole new set of tools to work with and a whole new context to work within. We went even further to say that perhaps many holonic things in life's holonarchy could be seen to occupy morphogenetic wavestates and to form a collective to which the above thermodynamics arguments apply. A cell for instance, could be seen to be the holonic entity, in place of a plant. The morphogenetic pattern state could be represented by its differentiation state, and thus, its presence in the body could be seen as its participation in a collective. Thus, the spontaneous nature of an organism's life cycle could be seen in terms of ever increasing emergatures in response to an emergent heat source (such as the eating of food).

Finally, we asked what these ideas might be used for, and proposed that they would assist most in a world where brute-force, oil-fed, mountain-moving machines could no longer grow food and extract resources from the Earth in order to maintain our enormous human population (Chapter 7). We discussed how at present our agricultural activities fight against the system's natural tendency to progress to its equilibrium emergature and confuse the emergent thermodynamics in a number of ways. We proposed that the growing of food without this kind of technological assistance would require a more profound understanding of the dynamics of plant communities so that we can learn how to actively progress their spontaneous natural tendencies in order to produce food and other resources for ourselves most effectively. We discussed permaculture as a viable alternative to mono-crop agriculture. In permaculture, there is the creation and cultivation of real plant communities involving a number of different elements that are self-supporting and grow in complexity and productivity each year. All of the plant elements should also have some utility for human beings, whether these are nutritional, medicinal, or structural properties. Emergent thermodynamics allows us to estimate the equilibrium emergature of the particular region we wish to grow food in by examining the state density diagrams of mature (100 years or greater) plant communities in the area, with a goal to develop a permaculture enterprise that ultimately reaches the equilibrium emergature of the area. Also, emergent thermodynamics shows us the relative proportion of various archetypal plant species at different levels of emergent energy during the natural life-cycle of a plant community. As emergent energy states are archetypal there are many different plant possibilities available at each emergent energy level. Thus, emergent thermodynamics helps to guide us in engineering harmonious plant communities by selecting plant elements that are of some benefit to humans but which also comply with the distribution required by the stage of the plant community's life cycle. However, in order to fully apply emergent thermodynamics for construction of food producing plant communities, further work is required. A thorough study of the distribution of natural plant communities (in emergent energy state density diagrams) at various ages of their development is required. Ultimately, human beings must strive to develop insulated emergent systems that are near equilibrium emergature as agricultural enterprises to provide food, medicine, and shelter in the self-interest of humanity. Unfortunately, none of this is a quick-fix solution, but requires a long-range view and cultivation spanning a few human generations.

Why should we care what science is and what it thinks about living things? Considering the advances of science in the past century, the technological fruits of science, and the way that these technologies have dramatically (and adversely) shaped and altered the living Earth, it is obvious that the actions and activities of our current culture are dominated by our rational knowings of reality. How we think is a direct mirror of the opportunities, problems and solutions we will perceive. Life-as-energy has been omitted from our rational vision, and therefore, it can be easily dismissed as something less real, less valuable, and with little consequence on the workings of the physical world. Moreover, omitting the concept of life-as-energy from our rational reasoning limits our capacity to understand natural systems. Consequentially, we are limited in our capacity to develop actions capable of integrating human interests with the ways that nature ultimately functions. It is hoped that this work will provide a new paradigm allowing for more effective understanding, valuation of, and sustainable integration of human activities with the natural world in which we are an inseparable element.

# Endnotes

**Chapter 1**

1. If your stomach is feeling strong, do consult the very detailed account of Descartes' machine view of living things and the corresponding scientific behaviours the animals-as-machines perspective led to in the text by Rupke 1987 pp. 24–30.
2. You can find a justification for the current legal requirement for the use of sedatives and painkillers during painful scientific experiments with animals in the Institute for Laboratory Animal Research (ILAR) documentation: ILAR 1996.
3. Here is a list of references detailing the various technological side-effects I mention: Risebrough *et al.* 1967 discuss DDT in food chains; the danger of commonly used chemicals accumulating in the environment is discussed by Bro-Rasmussen 1996; climate change by Houghton *et al.* 2001; and the unexpected consequences of stem cell therapies by Amariglio *et al.* 2009 and Canson *et al.* 2006.
4. In his book on Systems Science, Skyttner 2005, pp. 12–14 gives an excellent account of machine view thinking.
5. Great discussions about reductionist mindsets in the life science, including some of their limitations and consequences, can be found in the journal articles by Ahn *et al.* 2006, Peterson 2008, and Van Regenmortel 2004.
6. I got this figure of 100 trillion cells in the average human body from the journal article by Ahn *et al.* 2006.
7. Herman Haken is one of the world's great leaders in developing scientific theories which embody emergence. A discussion and definition of cyclic causality can be found in the journal article Haken *et al.* 1995.
8. This interchangeability of the parts is often called universality, and is discussed in the book by Sethna 2007, p. 191.
9. This is discussed in work that I did, which is also a discussion of using

whole thought systems from physics as conceptual metaphors for biological self-assembly processes. If interested, please check out my journal article, Pietak 2008.
10. In his book on Systems Theory, Skyttner 2005, pp. 36–47, does a great job discussing the systems thinking perspective.
11. This definition of complexity has a scientific basis which is explained in the journal article by West 2004.
12. Sornette 2000, pp. 240–50, does an excellent job comprehensively discussing what a spin system is and how its dynamics lead to critical phase transformations. The spin system is more technically known as the Ising Model and the Potts Model in physics.
13. This use of constructs similar to 'spin systems' to describe the foraging behaviour of ants can be found in the journal article by Beekman *et al.* 2001 and to describe earthquake predictions in Saleur *et al.* 1996.
14. This exciting study questioned a number of children from two different cultures and concluded that most children view life-as-energy before they get taught otherwise. This study is reported on in the journal article by Inagaki *et al.* 2004.
15. Scott Gilbert, a world renowned developmental biologist, is a proponent for more holistic scientific theories of living things. This, and the comment that more holistic theories may have been dismissed for political reasons, including their association with Nazi ideology, can be found in the journal article, Gilbert *et al.* 2000.

**Chapter 2**

1. The journal article by Sornette 1998 gives an excellent example of what symmetry of scale means.
2. One of my favourite books on the subject of fractals is by Petigen *et al.* 2003, and shows a great number of natural fractals on pages 242–47.
3. A great overview of the mathematical nature of various plant architectures is given in the journal article by Jean 1996.
4. I took this 'growth mechanism' example of a triangle growing on a square from Petigen *et al.* 2003, pp. 124–28.
5. The idea of creating chemical patterns as the route to creation of biological form is one of many theories of morphogenesis. Great descriptions of self-induced chemical patterning in morphogenesis are given in the journal articles by Koch *et al.* 1994, Meinhardt 1992, and Meinhardt 1996.
6. Morphogenesis is an immensely large and complex field and I've only

addressed the very tip of the morphic iceberg here. If you're interested in learning more, I can recommend the book by Gilbert 2006.

7. The 'multiple reduction copy machine' is a wonderful mechanism that's fully discussed in Peitgen *et al.* 2003, pp. 23–26 and pp. 217–37.

8. The amazing 'chaos game' is discussed in detail in Peitgen *et al.* 2003, pp. 34–36.

9. The idea of a fractal as an 'attractor pattern' was developed by Peitgen *et al.* 2003, pp. 218–38.

10. The less than glorious conclusion to the great genome sequencing projects is mentioned in detail in Ahn *et al.* 2006 and in Sheldrake 2009, pp. 45–50.

11. This very large figure for plant genes can be easily found in the scientific literature. I give the example of rice plant gene number here, taken from Sasaki *et al.* 2002.

12. The failure of the human gene sequencing project to deliver new medicines is discussed in the journal article by Ahn *et al.* 2006.

13. The wonderful illustration of the importance of context was given in Goodwin 2001, pp. 9–10.

**Chapter 3**

1. The nature and implications of conceptual metaphors and imaginative rationality are the surprisingly riveting subject of the book, *Metaphors We Live By*, by Lackoff and Johnson 1980.

2. This example, which considers conceptual metaphors for love, was taken directly from Lackoff and Johnson 1980, p. 49.

3. In this example, the metaphor of time-is-money was also discussed by Lackoff and Johnson 1980, p. 7.

4. The notion that mathematics is rooted in human experience was the subject of the remarkable book, *Where Mathematics Comes From*, by Lakoff and Nunez 2000. Inspired by their arguments that we invent our mathematics by configuring our basic human experiences, I am simply reformulating their arguments for physics instead of mathematics.

5. Check out any general physics or chemistry textbook to see how many systems are described in terms of the ball-on-a-hilly-landscape model for potential energy. For physics this might be Halliday *et al.* 1988 and for chemistry you might flip through Atkins 1994.

6. Imaginative rationality is not my own term. It has been defined and discussed previously, particularly by cognitive scientists like Lackoff and Johnson 1980, and Persmeg 1992.

7. If curious, check out Lovelock's first articles on the Gaia hypothesis in the journal article Lovelock 1972.
8. The details of the Gaia hypothesis, and the fact that it remains relatively excluded from mainstream science, are discussed by Skyttner 2005, pp. 140–52.
9. The morphogenetic field concept and its early inception by embryologists of the early twentieth century are discussed in the developmental biology book by Gilbert 2006 on pp. 67–68. Also, see the journal article by Beloussov 1997, as well as Weiss 1939, Harrison 1918, and Huxley 1934.
10. The work that Alexander Gurwitsch did on morphogenetic fields, as well as the trials that the concept came to face by the 1950s, are discussed in the journal article by Beloussov 1997.
11. Rupert Sheldrake discusses his key ideas in the book *A New Science of Life*, (Sheldrake 2009) as well as *The Sense of Being Stared at and Other Aspects of the Extended Mind* (Sheldrake 2003).

**Chapter 4**

1. The material presented in the next two sections ('Three modalities of matter and energy' and 'Life at the atomic level') represents a general-level synthesis of basic concepts from physics. My sources are the majority of the material from the general physics text Halliday and Resnick 1988, the book on vibrations and waves by French 1971, and the excellent quantum mechanics text by Shankar 1994.
2. David Bohm's pilot wave theory is outlined in the journal articles Bohm 1952a and Bohm 1952b. A further synthesis and discussion of his theory is given in the journal article Bohm 1990.
3. My favourite texts discussing the basic (and not so basic) ideas behind thermodynamics and statistical mechanics (entropy, heat, temperature, phase changes, and so on) are Sethna 2007 and Atkins 1994. Haken 1983, pp. 1–12, gives a great introduction to the basic concepts of entropy, and transitions between order and disorder (and vice versa).
4. One of the best descriptions for the 'reductionist' interpretation of temperature is given in the book by Sornette 2000, pp. 199–221.
5. An excellent discussion of exergy and how it can be used to account for the spontaneous appearance of order in complex systems is given in the journal article by Kay and Schneider 1994. The formation of order in dissipative open systems was initially the work of Prigogine and Stengers 1984.
6. I borrowed this example from Kay and Schneider 1994.

7. Haken discusses the multifoliated nature of reality and its implications in his book, *Information and Self-Organization*, Haken 2006, pp. 36–37.

**Chapter 5**

1. That pattern formation in leaves remains an unsolved problem is mentioned in Nelson *et al.* 1997.
2. If you are interested in the details of the ongoing search for the mechanisms behind pattern formation in leaves, see the following journal articles: Mitchison 1980, Klucking 1992, Uggla *et al.* 1996, Nelson *et al.* 1997, Swarp *et al.* 2000, Avsian-Kretchmer *et al.* 2002, Ye 2002, Scarpella *et al.* 2004, Feugier *et al.* 2005, Rolland-Lagan *et al.* 2005, Dimitrov *et al.* 2006; Feugier *et al.* 2006, Scarpella *et al.* 2006.
3. For the record, this is discussed by Gilbert in his developmental biology text, Gilbert 2006, as well as in a journal article, Gilbert 2000.
4. These next few sections on morphogenetic fields for leaves are my own work. Further details, including the mathematical formulations are discussed in the journal article, Pietak 2009.
5. This takes us back to David Bohm's work with the pilot wave interpretation of quantum mechanics, see Bohm 1952a, Bohm 1952b, Bohm 1990.
6. This amazing measurement of a very strong electric field inside cells is discussed in the journal article by Tyner *et al.* 2007.
7. This is my own calculation. Briefly, if we simplify the situation by saying the electric field, E, inside a cell is constant over space, then the energy stored in that field can be expressed:

$$Energy = (1/2) \times \varepsilon_r \times \varepsilon_o \times V \times E^2$$

Where (r is the relative permittivity of the body's electrolyte (taken here to be about 40), ($_o$ is the permittivity of free space (8.854 ×10–12 J/V2), V is the approximate volume of a cell (the cell is assumed to be a sphere with a radius of 5 (m), and E is the electric field strength inside the cell (where the average value of 2000000 V/m was taken). To get the energy stored in a human-sized organism, I multiplied the result by 100 trillion cells.
8. The fact that electric and magnetic fields are implicated in morphogenesis (including the example of the planaria field) is discussed in the excellent review articles by Nuccitelli 1992 and Levin 2003.
9. Knowledge of electric fields existing in a wide variety of organisms and their correlation to a number of biological and environmental processes

emerged primarily from the highly prolific research activity of a scientist named Harold Burr and his students Ravitz and Gott, working from about 1940–1960. Electric fields were measured in trees (Burr 1947), squash fruits (Burr and Sinnot 1944), rats (Burr 1940), monkeys (Gott and Burr 1953), humans (Ravitz 1951). They have been correlated to growth processes (Burr 1942, Burr and Sinnot 1944), to ovulation events in monkeys and humans (Gott and Burr 1953), to daily and seasonal rhythms (Burr 1947, Ravitz 1951, Leach 1987) and are even indicative of cancer (Burr 1940 and Burr 1941). For unknown reasons, the abundant research on electric fields has remained almost completely ignored by scientists since. Although more recent reports with modern equipment (such as Leach 1987) confirm the existence of effects measured by Burr.

10. The possibility that endogenous electromagnetic fields can form in water and in biological systems has been discussed by a number of researchers. Frohlich 1968a, 1968b and 1975, was the first to propose that the vibration of polar molecules (like water or a variety of cell structures) could lead to coherent state and a self-trapped electromagnetic field within the substance. Further theoretical work exploring the formation of endogenous electromagnetic fields in water, electrolytes, and biological systems in terms of quantum field theory was completed by Arani *et al.* 1995, and more recently by Del Giudice *et al.* 2000, Del Giudice *et al.* 2005, and Pokorny *et al.* 2005.

11. Brizhik *et al.* 2009b and Del Giudice *et al.* 2009 have both proposed that endogenous electromagnetic fields could span ranges from cells, organs, whole life forms, and even ecosystems.

12. Jerman *et al.* 2009 have written a beautiful article discussing the feasible relationship between endogenous electromagnetic fields arising from vibrating charged biological bits, and the concept of the morphogenetic field.

13. Pokorny *et al.* 2001 and Cifra *et al.* 2008 have detected radio-waves from yeast cultures. Microwave emission from frog legs was detected by Gebbie and Miller in 1997. Light emission from living things is a very well characterized phenomenon reviewed in: Van Wijk 2001, Choi *et al.* 2002, Popp *et al.* 2002a, Popp *et al.* 2002b, Jung *et al.* 2005, Popp *et al.* 2005, and Popp *et al.* 2009.

14. Alexander Gurwitsch's early work with 'mitogenic radiation' is discussed in Van Wijk 2001.

15. A number of research articles explore 'ultraweak light' or 'biophoton' emission from living organisms. For example, we have the journal articles: Van Wijk 2001, Choi *et al.* 2002, Popp *et al.* 2002a, Popp *et al.* 2002b, Jung *et al.* 2005, Popp *et al.* 2005, and Popp *et al.* 2009.

16. On the subject of light emission being used to coordinate growth processes in a collective of cells, see the review article by Van Wijk 2001.
17. Research that has examined the very interesting coherence properties of light emission from living organisms can be found in the work of Fritz Albert Popp in the journal articles Popp *et al.* 2002a, Popp *et al.* 2002b, Popp *et al.* 2005, and Popp *et al.* 2009.

**Chapter 6**
1. This is discussed by a number of ecologists and is mentioned in the journal article by Kay and Schneider 1994.
2. A discussion of natural succession can be found in any ecology text such as Sole and Bascompte 2006. See also the journal article by Connell and Slayter 1977.
3. The notion that ecological network analysis is the closest thing to a theoretical framework that ecology has, was expressed by Zavaleta 2004. Professor Robert Ulanowicz is the world's leading expert on ecological network analysis. If you are interested in learning more, I recommend his review paper, Ulanowicz 2004.

**Chapter 7**
1. I found this fact on fossil fuel usage in electricity in the journal article by Dukes 2003.
2. These amazing (and horrifying) estimates of the plant material in gasoline are from the journal article by Dukes 2003.
3. The environmental facts that follow this statement were taken from the journal article by Kay and Schneider 1994.
4. This definition occurs in the Preface of Mollison's 1997 book, *Permaculture: a Designer's Manual.*
5. To learn more about calorific restriction (fasting) and lifetime extension in all organisms studied, see the journal article by Koubova and Guanente 2003, as well as Weindruch *et al.* 1986.

# List of Captions

Figure 1.1: The primary phases of matter (solid, liquid, and gas) can be explained by understanding the relationships between their elemental parts (atoms or molecules).

Figure 1.2: A scientific system showing the boundary, the environment, and the idea of system inputs and outputs.

Figure 1.3: Examples of simple and complex patterns. A highly ordered, featureless pattern like a solid triangle is simple as it's easy to describe (panel A). However, a random pattern like static on a television is also considered simple, as it's also easy to describe (panel B). A truly complex pattern contains intricate features and a lot more work is necessary to come up with a description for it (panel C).

Figure 1.4: These arrows represent the tiny atoms called spins present in a piece of magnetic material like iron. If two spins are placed side by side, they are most comfortable when pointing in the same direction as the energy of their interaction is minimized. However, if all spins point in the same direction, the disorder (entropy) of the system would also be spontaneously minimized, which is unfavourable. Therefore these two spin configurations are associated with conflicting drives: that of minimizing energy and that of maximizing entropy.

Figure 1.5: The whole material is macroscopically magnetic only when all of the spins are aligned in the same direction (left). This aligned state corresponds to minimizing the energy of the system. When the spins face random directions (right), there is no magnetism in the whole. This randomized state corresponds to maximizing the entropy of the system. The temperature of the system determines which state will exist. Low temperatures favour alignment and minimized energy; high temperatures favour randomization and maximized entropy.

Figure 2.1: The concept of hierarchical structure.

Figure 2.2: Symmetries of a square (top) and a waveform (bottom).

Figure 2.3: The Sierpinski triangle fractal.

the particles passed through the slits.

Figure 4.20: The interference pattern for particle-like electrons passing through two closely spaced slits. The pattern does not form gradually and evenly, but as discrete units appearing on the screen.

Figure 4.21: In the interpretation of Bohmian mechanics the electrons remain particles and are carried into an interference pattern on the back of a pilot wave.

Figure 4.22: A system-of-systems.

Figure 4.23: The primordial paradigm of discrete and continuous.

Figure 4.24: The idea of 'state' as a position in space is extended to assume many different meanings.

Figure 4.25: Possible energy states (top), state diagram (middle), and state density diagram (bottom) for a quantum system-of-systems.

Figure 4.26: Entropy is all about the arrangement of items in states, whether they be position states (as shown in the figure) or other kinds of states (like energy states). The top image shows order and low entropy, the bottom image shows disorder and high entropy.

Figure 4.27: Without coherent energy input, systems-of-systems are driven to proceed from order to disorder, or, in other words, from low to high entropy.

Figure 4.28: Non-living objects like a manmade pair of pliers (top) spontaneously rust, returning to a disorganized, high entropy state. Living objects like the sprout spontaneously grow into intricate, complex forms suggesting an apparent decrease in entropy.

Figure 4.29: Temperature corresponds to the distribution of a collection of items in their energy states. Here a state diagram illustrates the meaning of low and high temperatures.

Figure 4.30: Here a state density diagram illustrates the meaning of low and high temperatures in terms of the distribution of particles in their energy states.

Figure 4.31: A dissipative open system where matter and energy flow in and out.

Figure 4.32: The 'tornado in a bottle' toy. Water in the top flows more rapidly to the bottom jug in the form of a swirling vortex.

Figure 4.33: Science, and its perspectives of the world, are foliated as the microscopic, mesoscopic, and macroscopic. Different scientific systems of thought have been developed for each of these levels of scale.

Figure 5.1: A leaf taken from the bush where I sat eating my lunch.

Figure 5.2: Archetypal leaf patterns. In order, from left to right, are shown examples of needle, blade, fishbone, and rotator structures.

# Glossary

*Algorithm* ~ A procedure, rule or formula for solving a problem in a finite number of steps. The term derives from the name of an Arab mathematician, Al-Khawrizm.

*Amplitude* ~ Pertaining to a periodic function, it is the maximum displacement from a reference level in either a positive or negative direction. The periodic function can represent a vibration or wave, or can describe the motion of a point on a pendulum or spring.

*Auxin* ~ A protein plant hormone implicated in many biological processes, especially growth and development of the plant.

*Biomass* ~ The weight of living matter and organic material (wood, agricultural wastes).

*Coherent* ~ Sticking together; orderly, logical and consistent.

*Complexity* ~ In mathematics, complexity refers to the number of discrete steps (such as addition or multiplication) needed to complete an algorithm, expressed as a function of the input size. In Organic Mechanics complexity refers to how many hierarchical levels are involved in the generation of a particular structure, as more hierarchical levels of structure generates more involved and detailed patterning of the form.

*Conceptual metaphor* ~ The understanding of one idea or conceptual domain in terms of another. A conceptual domain can be any coherent organization of human experience.

*Continuous* ~ An entity, substance or process whose structure, distribution or properties are unbroken.

*Co-ordinate* ~ A number specifying the position of a point relative to certain other lines or points.

*Degeneracy* ~ Pertaining to quantum mechanics, in a system of wavefunctions, this is the existence of states with different forms that share the same energy value.

*Discrete* ~ In an entity, substance or process, there are regions of structure, distribution or property, between which the entity, substance or process does not exist.

*Emergent energy* ~ A form of energy proposed to have unique effects and properties; however, depends on the interactions and relationships of a supporting system.

*Emergent property* ~ A property or function that arises out of interactions and relationships amongst the parts of a collective system and is not present in any individual part of the collective.

*Emergent phase* ~ A patterned organisation of a collective system which supports an emergent property.

*Emergature* ~ Pertains to the dispersion of elements in emergent energy states.

*Emergent heat* ~ Sources of energy, nutrients or conditions which are conducive to the growth and development of a living system. The sources of emergent heat are relevant to the needs of the specific system.

*Empirical* ~ That which is based on observation or experiment rather than deduced from basic laws or postulates.

*Entropy* ~ A measure of the amount of disorganisation or possibility.

*Exergy* ~ The quality of an energy form. Higher exergy energy is able to do more work than lower exergy energy.

*Feedback* ~ The process in which part of the output of a system is returned to its input in order to regulate further output.

*Frequency* ~ Referring to a periodic function, this is the number of complete oscillations or cycles that occur in a unit of time or space reflecting the rate of repetition of a periodic phenomenon.

*Harmonics* ~ Vibrating objects usually have a main fundamental frequency. This fundamental frequency has associated with it a series of other frequencies which are integer multiples of the fundamental called the harmonics.

*Hierarchical* ~ A logical or physical structure classifying information or structure in a series of steps, starting with broad, simple classifications or structures and proceeding in stages to narrow precise classifications or structures.

*Holon* ~ Something which is simultaneously a semi-autonomous entity, and part of a larger whole.

*Holonarchy* ~ A hierarchy made up of holons.

*Individuated* ~ To give an individual or distinctive character to; to form into an individual or distinct entity.

*Inorganic* ~ Matter other than plant or animal and not containing a combination of Carbon, Hydrogen or Oxygen, as in living things.

*Kinetic energy* ~ The energy of motion.

*Matter* ~ That which has mass and occupies space.

*Mechanistic* ~ As in a machine, this describes a process that runs on a formalized, axiomatic system or algorithm.

*Mono-crop* ~ To grow only one plant type in a region, usually as an agricultural enterprise. For example, a field of wheat.

*Orthogonal* ~ At right angles to one another, as in the corners of a perfect square.

*Organic* ~ Being, or relating to, or derived from, or having properties characteristic of living organisms.

*Organism* ~ A living being.

*Organizing region* ~ In biological development, this is a localized, specialized region of cells that assists in creating the next level of biological pattern.

*Potential energy* ~ Energy stored within a system. This energy can be released or converted into other forms of energy.

*Quantized* ~ A physical parameter which can only take on certain discrete values.

*Reductionalist* ~ To decompose into pieces that are fundamental elements which are thought to account for the properties of the whole.

*Systematic* ~ Characterized by order and planning.

*Synergistic* ~ The simultaneous action of separate things, which has a greater total effect than the sum of their individual effects.

*Temperature* ~ Pertaining to the dispersion of elements in kinetic energy states.

*Thermodynamics* ~ The physical theory of heat and energy distribution in the universe.

*Wavelength* ~ A property of a wave, expressed as the distance travelled in the direction of propagation between two points at the same phase of disturbance in consecutive cycles of the wave.

*Work* ~ That which is not spontaneous. A coherent, directed energy input. The application of force to transfer kinetic energy to an object.

# Bibliography

Ahn, A., Tewari, M., Poon, C-S. & Phillips, R. (2006) The limits of reductionalism in medicine: Could systems biology offer an alternative? *Public Library of Science (PLoS) Medicine* 3:0700–0713.

Amariglio, N., Hirshberg, A., Scheithauer, B.W., Cohen, Y., Loewenthal, R., Trakhtenbrot, L., Paz, N., Koren-Michowitz, M., Waldman, D., Leider-Trejo, L., Toren, A., Constantini, S. & Rechavi, G. (2009) Donor-Derived Brain Tumor Following Neural Stem Cell Transplantation in an Ataxia Telangiectasia Patient. *PLoS Med* 6:e1000029.

Arani, R., Bono, I., Del Giudice, E. & Preparata, G. (1995) QED coherence and the thermodynamics of water. *Int J of Modern Physics B* 9:1813–1841.

Atkins, P. (1994) *Physical Chemistry, 5th Edition.* W.H. Freeman Press, Toronto.

Avsian-Kretchmer, O., Cheng, J., Chen, L., Moctezuma, E. and Sung, Z. (2002) Indole acetic acid distribution coincides with vascular differentiation pattern during arabidopsis leaf ontogeny. *Plant Physiology* 130:199–209.

Beekman, M., Sumpter, D. & Ratnieks, F. (2001) Phase Transition Between Disordered and Ordered Foraging in Pharaoh's Ants. *Proceedings of the National Academy of Sciences (PNAS)* 98:9703–9706.

Bell, J.S. (1987) *Speakable and Unspeakable in Quantum Mechanics.* Cambridge University Press, Cambridge, UK.

Beloussov, L. (1997) Contributions to field theory and life of Alexander G. Gurwitsch. *Int. Journal of Developmental Biology* 41:771–779.

Bohm, D. (1952a) A suggested interpretation of quantum theory in terms of "hidden" variables I. *Physical Review* 85:166–179.

—, (1952b) A suggested interpretation of quantum physics in terms of "hidden" variables. II. *Physical Review* 85:180–193.

—, (1990) A new theory of the relationship of mind and matter. *Philosophical Psychology* 3:271–286.

Brizhik L., Del Giudice, E., Popp, F., Maric-Oehler, W. & Schlebusch,

K. (2009) On the dynamics of self-organization in living organisms. *Electromagnetic Biology and Medicine* 28:28–40.
Brizhik L., Del Giudice E., Jorgensen, S., Marchettini, N., & Tiezzi, E. (2009b) The role of electromagnetic potentials in the evolutionary dynamics of ecosystems. *Ecological Modelling.* 220:1865–9.
Bro-Rasmussen, F. (1996) Contamination by persistent chemicals in food chains and human health. *Science Tot Env* 188:545–560.
Burr, H. (1940) Electrometric studies of tumors in mice induced by the external application of benzpyrene. *Yale Journal of Biology and Medicine* 12(6):711–7.
—, (1941) Changes in the field properties of mice with transplanted tumors. *Yale J Biol Medicine* 13(6):783–8.
—, (1942) Electrical correlates of growth in corn roots. *Yale J Biol Medicine* 4(6):581–8.
—, (1947) Tree potentials. *Yale J Biol Med.* 19(3):311–8.
Burr, H. & Sinnott, E. (1944) Electrical correlates of form in cucurbit fruits. *American J Botany.* 31(5):249–53.
Canson, C., Aigner, S. & Gage, F. (2006) Stem cells: the good, the bad, and the barely in control. *Nature Medicine* 12:1237–1238.
Choi, C., Woo, W., Lee, M., Yang, J. & Soh, K. (2002) Biophoton emission from the hands. *J Korean Physical Society* 41:275–278.
Cifra, M., Pokorny, J., Jelinek, F., & Hasek, J. (2008) Measurement of yeast cell electrical oscillations around 1 kHz. *Progress in Electromagnetic Research Symposium (PIERS) Proceedings.* July 2–6, 2008; Cambridge, USA, pp. 780–85.
Connell, J. & Slayter, R. (1977) Mechanisms of succession in natural communities and their role in community stability and organization. *American Naturalist* 111:1119–1144.
Del Giudice, E., Pulselli, R., & Tiezzi, E. (2009) Thermodynamics of irreversible processes and quantum field theory: an interplay for the understanding of ecosystem dynamics. *Ecological Modelling.* 220(16):1874–9.
Del Giudice, E., De Ninno, A., Fleischmann, M., Mengoli, G., Milani, M., Talpo, G. & Vitiello, G.. (2005) Coherent quantum electrodynamics in living matter. *Electromagnetic Biology and Medicine* 24:199–210.
Del Giudice, E., Doglia, S. & Milani, M. (1985) Quantum field theoretical approach to the collective behaviour of biological systems. *Nuclear Physics B* B251:375–400.
Del Giudice, E., Preparata, G. & Fleischmann, M. (2000) QED coherence and electrolyte solutions. *J Electroanalytical Chemistry* 482:110–116.
Dimitrov, P. & Zucker, S. (2006) A constant production hypothesis guides

leaf venation patterning. *Proceedings of the National Academy of Sciences (PNAS)* 103:9363–9368.

Dukes, J. (2003) Burning buried sunlight: Human consumption of ancient solar energy. *J Climate Change* 61:31–34.

Estes, C. (1995) *Women Who Run With the Wolves: Myths and Stories of the Wild Woman Archetype*. Ballantine Books, New York.

Feugier, F. & Iwasa, Y. (2006) How canalization can make loops: A new model of reticulated leaf vascular pattern formation. *J Theoretical Biology* 243:235–244.

Feugier, F., Mochizuki, A. & Iwasa, Y. (2005) Self–organization of the vascular system in plant leaves: Inter-dependent dynamics of auxin flux and carrier proteins. *J Theoretical Biology* 236:366–375.

French, A. (1971) *Vibrations and Waves*. W.W. Norton and Co., New York.

Frohlich, H. (1968a) Bose condensation of strongly excited longitudinal electric modes. *Physics Letters A* 26A:402–403.

—, (1968b) Long-range coherence and energy storage in biological systems. *International J of Quantum Chemistry* 2:641–649.

—, (1975) Evidence for bose condensation-like excitation of coherent modes in biological systems. *Physics Letters A* 51A:21–22.

Gebbie, H., & Miller, P. (1997) Nonthermal microwave emission from frog muscles. *Int J Infrared and Millimeter Waves*. 18(5):951–7.

Gilbert, S. & Sarkar, S. (2000) Embracing complexity: Organicism for the 21st century. *Devel Dyn* 219:1–9.

Gilbert, S. (2006) *Developmental Biology*. Sinauer Associates, New York.

Goodwin, B. (2001) *How the Leopard Changed Its Spots: The Evolution of Complexity*. Princeton University Press, Princeton.

Gott, V. & Burr, H. (1953) Electrical correlates of ovulation in the Rhesus monkey. *Yale J Biol Medicine* 25(5):408–17.

Haken, H. (1983) *Synergetics: An Introduction 3rd Edition*. New York.

—, (2006) *Information and Self-Organization: A Macroscopic Approach to Complex Systems, 3rd Ed*. Springer-Berlin-Heidelberg London.

Haken, H., Wunderlin, A. & Yigitbasi, S. (1995) An introduction to synergetics. *Open Systems and Information Dynamics* 3:97–130.

Halliday, D. & Resnick, R.(1988) *Fundamentals of Physics*. Wiley, Toronto.

Harrison, R. (1918) Experiments on the development of the fore-limb of Amblystma, a self-differentiating equipotential system. *Journal of Experimental Zoology* 25:413–461.

Houghton, J., Ding, Y., Griggs, D., Noguen, M., van der Linden, P., Dai, X., Maskell, K. & Johnson, C. (2001) *Climate Change 2001: The Scientific Basis*. Cambridge University Press.

Huxley, J. & de Beer, G. (1934) *The Elements of Experimental Embryology.* Cambridge University Press, Cambridge.

ILAR (1996). *Guide for the care and use of laboratory animals.* CoL Sciences. Washington, DC, National Academies Press.

Inagaki, K. & Hatano, G. (2004) Vitalistic causality in young children's naive biology. *Trends in Cognitive Sciences* 8:356–362.

Jean, R. (1996) Phyllotaxis: The status of the field. *Mathematical Biosciences* 127:181–206

Jerman, I., Krasovec, R., & Leskovar, R. (2009). Deep significance of the field concept in contemporary biomedical sciences. *Electromagnetic Biology and Medicine* 28:61–70.

Jung, H., Yank, J., Woo, W., Choi, C., Yang, J. & Soh, K. (2005) Year-long biophoton measurements: normalized frequency count analysis and seasonal dependency. *J Photochemistry and Photobiology B* 78:149–154.

Kay, J. & Schneider, E. (1994) Complexity and thermodynamics: Towards a new ecology. *Futures* 26:626–647.

Klucking, E. (1992) *Leaf Venation Patterns.* J. Cramer, Berlin.

Koch, A. & Meinhardt, H. (1994) Biological pattern formation: from basic mechanisms to complex structures. *Reviews of Modern Physics* 66:1481–1494

Koubova, J. & Guanente, L. (2003) How does calorie restriction work? *Genes and Development* 17:313–321.

Lakoff, G. and Johnson, M. (1980) *Metaphors We Live By.* University of Chicago Press, London.

Lakoff, G. & Nunez, R (2000) *Where Mathematics Comes From: How the Embodied Mind Brings Mathematics into Being.* Basic Books, New York.

Leach, C. (1987) Diurnal electrical potentials of plant leaves under natural conditions. *Environmental and Experimental Botany* 27(4):419–30.

Levin, M. (2003) Bioelectromagnetics in Morphogenesis. *Bioelectromagnetics* 24:295–315.

Lovelock, J. (1972) Atmospheric homeostasis by and for the biosphere — the Gaia hypothesis. *Tellus* 26:2-10.

Meinhardt, H. (1992) Pattern formation in biology: A comparison of models and experiments. *Reports on Progress in Physics* 55:797–849

—, (1996) Models of biological pattern formation: common mechanisms in plant and animal development. *Int J Dev Biol* 40:123–134.

Milani, M., Del Giudice, E.., Santisi, G., Talpo, G. & Vitiello, G. (2005) Yeast suspensions: a controllable example of a coherent quantum machine? *Electromagnetic Biology and Medicine* 24:331–340.

Mitchison, T. (1980) A model for vein formation in higher plants. *Proceedings*

*of the Royal Society London B — Biological Sciences* 207:79–109.
Mollison, B. (1997) *Permaculture: A Designer's Manual.* Tagari Publications, Sydney.
Nelson, T. and Dengler, N. (1997) Leaf vascular pattern formation. *Plant Cell* 9:1121–1135.
Nuccitelli, R (1992) Endogenous ionic currents and DC electric fields in multicellular animal tissues. *Bioelectromagnetics* S1:147–157.
Persmeg, N. (1992) Prototypes, metaphors, metonymics and imaginative rationality in high school mathematics. *Educational Studies Mathematics* 23:595–610.
Peterson, R (2008) Chemical biology and the limits of reductionalism. *Nature Chemical Biology* 4:635–638.
Peitgen, H-O., Jurgens, H. & Saupe, D. (2003) *Chaos and fractals: New Frontiers of Science. 2nd Edition*. Springer, New York.
Pietak, A. (2008) Seeing Tissue as a Phase of Matter. *Physical Biology* 5: 016007–016010.
—, (2009) Describing long-range patterns in leaf vasculature with metaphoric fields. *J Theoretical Biology* 261:279–289.
Pokorny, J., Hasek, J. & Jelinek, F. (2005) Endogenous electric field and organization of living matter. *Electromagnetic Biology and Medicine* 24:185–197.
Pokorny, J., Hasek, J., Jelinek, F., Saroch, J., & Palan, B. (2001). Electromagnetic activity of yeast cells in the M phase. *Electro and Magnetobiology.* 20(3):371–96.
Popp, F. (2009) Cancer growth and its inhibition in terms of coherence. *Electromagnetic Biology and Medicine* 28:53–60.
Popp, F., Chang, J., Herzog, A., Yan, Z. and Yan, Y. (2002) Evidence of non-classical (squeezed) light in biological systems. *Physics Letters A* 293:98–102.
Popp, F., Sigrist, S., Schlesinger, D., Dolf, A., Yan, Y., Cohen, S. & Chotia, A. (2005) Further analysis of delayed luminescence of plants. *J Photochemistry and Photobiology B* 78:235–244.
Popp, F. & Yan, Y. (2002) Delayed luminescence of biological systems in terms of coherent states. *Physics Letters A* 293:93–97.
Prigogine, I. & Stengers, E. (1984) *Order Out of Chaos.* Bantam Books, New York.
Ravitz, L. (1951) Daily variations of standing potential differences in human subjects. *Yale J Biol Medicine* 24(1):22–25.
Risebrough, R., Menzel, D., Jun, J. & Olcott, H. (1967) DDT residues in Pacific sea birds: A persistent insecticide in marine food chains. *Nature*

216:589–591.

Rolland-Lagan, A. & Prusinkiewicz, P. (2005) Reviewing models of auxin canalization in the context of leaf vein pattern formation in Arabidopsis. *The Plant J* 44:854–865.

Rupke, N. (1987) *Vivisection in Historical Perspective.* Croom Helm, London.

Saleur, H., Sammis, C. & Sornette, D. (1996) Discrete scale invariance, complex fractal dimensions, and log-periodic fluctuations in seismicity. *J Geophys Res* 101:17,661–617,677.

Sasaki, T. (2002) The genome sequence and structure of rice chromosome I *Nature* 420:312–316.

Scarpella, E., Marcos, D., Triml, J. & Berleth, T. (2006) Control of leaf vascular patterning by polar auxin transport. *Genes and Development* 20:1015–1027.

Scarpella, E. & Meijer, A. (2004) Pattern formation in the vascular system of monocot and dicot plant species. *New Phytologist* 164:209–242.

Sethna, J. (2007) *Statistical Mechanics: Entropy, Order Parameters and Complexity.* Clarendon Press, Oxford.

Shankar, R. (1994) *Principles of Quantum Mechanics. 2nd Edition.* Springer, New York.

Sheldrake, R. (2003) *The Sense of Being Stared at and Other Aspects of the Extended Mind.* Arrow Books, London.

—, (2009) *A New Science of Life: The hypothesis of formative causation.* Icon Books, London.

Skyttner, L. (2005) *General Systems Theory: Problems, Perspectives and Practice.* World Scientific, New Jersey.

Sole, R. & Bascompte, J. (2006) *Self-Organization in Complex Ecosystems.* Princeton University Press, Princeton.

Sornette, D. (1998) Discrete scale invariance and complex dimensions. *Physics Reports* 297:239–270.

—, (2000) *Critical Phenomena in Natural Sciences: Chaos, fractals, self-organization and disorder: Concepts and tools.* Springer, New York.

Swarup, R., Marchant, A. & Bennett, M.J. (2000) Auxin transport: Providing a sense of direction during plant development. *Biochemical Society Transactions* 28:481–485.

Tyner, K., Kopelman, R. & Philbert, M. (2007) "Nanosized Voltmeter" enables cellular-wide electric field mapping. *Biophysical J* 93:1163–1174.

Uggla, C., Moritz, T., Sandberg, G. & Sundberg, B. (1996) Auxin as a positional signal in pattern formation in plants. *Proceedings of the National Academy of Sciences* 93:9282–9286.

Ulanowicz, R. (2004) Quantitative methods for ecological network analysis.

*Computational Biology and Chemistry* 28:321–339.

Van Regenmortel, M. (2004) Reductionalism and complexity in molecular biology. *European Molecular Biology Organization (EMBO) Reports* 4:1016–1020.

Van Wijk, R. (2001) Bio-photons and bio-communication. *J Scientific Exploration* 15:183–197.

Weindruch, R., Walford, R., Fligier, S. & Guthrie, D. (1986) The retardation of aging in mice by dietary restriction: Longevity, cancer, immunity and lifetime energy intake. *J of Nutrition* 116:641–654.

West, B. (2004) Comments on the renormalization group, scaling and measures of complexity. *Chaos, Solitons and Fractals* 20:33–34.

Ye, Z. (2002) Vascular tissue differentiation and pattern formation in plants. *Annual Review of Plant Biology* 53:183–202.

Zavaleta, E., Hobbs, R. & Mooney, H. (2001) Viewing invasive species removal in a whole-ecosystem context. *TRENDS in Ecology and Evolution* 16:454–459.

# Index